Henry C. Burdett

Cottage Hospitals, General, Fever and Convalescent

Their progress, management and work in Great Britain and Ireland, and the United

States of America: with an alphabetical list of every cottage hospital at present

opened

Henry C. Burdett

Cottage Hospitals, General, Fever and Convalescent
Their progress, management and work in Great Britain and Ireland, and the United States of America: with an alphabetical list of every cottage hospital at present opened

ISBN/EAN: 9783337325732

Printed in Europe, USA, Canada, Australia, Japan

Cover: Foto ©berggeist007 / pixelio.de

More available books at **www.hansebooks.com**

COTTAGE HOSPITALS

General, Fever, and Convalescent

THEIR PROGRESS, MANAGEMENT, AND WORK IN GREAT BRITAIN AND IRELAND, AND THE UNITED STATES OF AMERICA

WITH AN ALPHABETICAL LIST OF EVERY COTTAGE HOSPITAL AT PRESENT OPENED

MANY PLANS AND ILLUSTRATIONS

INCLUDING A

PORTRAIT OF ALBERT NAPPER, ESQ.
The Founder of Cottage Hospitals

BY

HENRY C. BURDETT

AUTHOR OF "HOSPITALS AND ASYLUMS OF THE WORLD"; "BURDETT'S HOSPITALS
AND CHARITIES: BEING THE YEAR BOOK OF PHILANTHROPY;" "HOSPITALS
AND THE STATE ;" "PAY HOSPITALS OF THE WORLD ;" "THE RELATIVE
MORTALITY OF LARGE AND SMALL HOSPITALS," AND EDITOR OF
"THE HOSPITAL."

Third Edition. Rewritten and much New Matter Added

LONDON
THE SCIENTIFIC PRESS, Limited, 428 Strand, W.C.
1896

PREFACE TO THE THIRD EDITION.

IT is fifteen years since the Second Edition of this book was published. During the last few years a Third Edition has been in urgent demand; but owing to the great pressure of other work we have found it impossible to produce a new edition earlier. It was also felt that it would be better for some one else to produce a book on the subject, than to merely reprint the Second Edition without adequate revision. Now, however, the present volume is not only up to date, but is in reality a new book on the subject of cottage hospitals.

Since the First Edition was published, nearly twenty years ago, it has been our privilege and pleasure to do our utmost to promote the success of cottage hospitals everywhere. No occasion has been missed to insist upon their value and importance to all who reside in rural districts. It has been urged that the cottage hospital enabled the country practitioner to become an accomplished and skilful surgeon, and to keep himself well abreast of the new features of medical practice.

(v)

In God's good providence we have had evidence
of the precise truth of our contentions in the person
of one of our sons. This schoolboy, during the
recent general election, whilst engaged in dragging
the carriage of the member for the division round the
quad., at Marlborough College, was thrown down
with others, and received very serious injuries, which,
in the absence of skilled medical treatment, must
have led to permanent disfigurement, if not to death.
Fortunately, Dr. J. B. Maurice and his son Dr. Thurl-
wall Maurice, members of the staff of the Savernake
Cottage Hospital, were within call at the time of the
accident. By their devotion and skill the lad's life
was spared, and the worst effects of his injuries were
successfully removed. It is seldom that so painful
and providential an example of the value of an
author's teaching is brought home to him during
his lifetime. In the case of this Marlborough
boy, the grateful father feels that any indebtedness
and obligation on the part of cottage hospitals
and their medical staffs and managers to himself,
have been entirely counterbalanced by the magnificent services rendered to his dear son by the surgeons
of the Savernake Cottage Hospital.

Up to this time it has been customary for the Prime
Minister, or whoever has had to advise the Queen
in matters of this kind, to confine the honours given
to members of the medical profession almost

entirely to the physicians or surgeons of a few
large hospitals. We are strongly of opinion
that the time has arrived when the great services
rendered to millions of the population throughout
the United Kingdom by the medical officers of
cottage hospitals should be recognised and acknowledged by the State. We are confident that the
matter has only to be brought to the notice of
Her Majesty to ensure that it will be dealt with
entirely on its merits, and that justice will be
done to the great body of medical practitioners
attached to cottage hospitals, who have probably
done as much for their day and generation as some
of their better known *confrères* who are attached to
the larger hospitals in great cities and the Metropolis.

The Second Edition of this book had two faults.
It was too large and too expensive. With a view to
remedying these defects, we have omitted the special
chapters on midwifery in cottage hospitals, cottage
hospital mortuaries, and convalescent cottages,
which were given in the Second Edition, though we
have found space to deal with all these questions
briefly but adequately under their proper heads in
other chapters of the present book.

The Second Edition also contained several appendices, occupying nearly 100 pages, now omitted.
In their place we have given an alphabetical list of
all the cottage hospitals in the United Kingdom

and in the United States of America of which we have any knowledge.

In response to many requests we have added a model set of rules for the guidance of cottage hospital managers. These rules are in effect a codification of the best features of the rules in force at existing hospitals, and we hope they may be found of service in many ways.

In the present volume a great deal of attention has been bestowed on the chapters on construction and plans. We have added, for instance, a series of model plans on the straight as opposed to the pavilion system, because experience has shown that the pavilion is in practice too costly a system for so small an institution as a cottage hospital proper. Altogether we have inserted in the present edition about double the number of plans formerly published, in the belief that such information is greatly desired by all who are interested in cottage hospitals, or who contemplate taking part in the establishment of a new institution.

Finally, attention is specially drawn to the new and fairly exhaustive section dealing with cottage hospitals in the United States of America. In the smaller towns of New England the cottage hospital system has made rapid progress during the last fifteen years. There is evidence, too, that many other States are following New England in this matter, to the no small advantage and benefit of the rural or semi-urban populations.

We have again the pleasant duty of recognising the hearty co-operation and assistance afforded us by the medical officers and honorary secretaries of the various institutions with whom we have had cause to communicate. When the previous editions were issued, the managers of many cottage hospitals helped to make the book known by publishing its title page on the last page of their Annual Reports. It was a good idea, and proved helpful in practice to the institutions in question, by inducing their supporters to buy the book and take a more intelligent and active interest in the local cottage hospitals.

LONDON, *January.* 1896.

EXTRACTS FROM PREFACE TO THE SECOND EDITION.

"IN CHARACTER, IN MANNER, IN STYLE, IN ALL THINGS, THE SUPREME EXCELLENCE IS SIMPLICITY."– *Longfellow.*

THE satisfaction with which the author naturally presents a Second Edition of this book for the assistance and counsel of those who are interested in the smaller medical charities of all descriptions, —fever and acute hospitals, free and provident dispensaries, and convalescent institutions,—is tempered by a feeling of regret that so few cottage hospital managers seem to have tried to secure that uniformity of management and book-keeping which are so desirable. Apart from this everything is satisfactory. Numerous new cottage hospitals have been opened, many more have been built, and nearly all have increased in efficiency. Every day the public and the profession are placing a higher value on the work done at the cottage hospital. It secures to the former the immediate service of a skilled and practised medical attendant, who was formerly rarely obtainable, except in the larger centres of population,

(x)

and it enables the latter to have a wider field for
practice under conditions which reward care and skill,
whilst they increase both reputation and income.
To recognise these facts, is to admit that the cottage
hospital has now established itself firmly in public
estimation, to the great good of all concerned.

.   .   .   .   .   .

The author has reason to think there are good
grounds for believing that the cottage hospital will
soon prove as popular in the United States of
America as it has become in Great Britain.

.   .   .   .

An excellent engraving, representing Mr. Albert
Napper, the founder of cottage hospitals, with a
sketch of the first Village Hospital, opened at Cranleigh in 1859, forms the frontispiece of this edition.
The engraving is a fairly faithful likeness, and all
who know Mr. Napper will value and appreciate
this addition to the book for his sake.

.   .   .   .

To the many reviewers, English and foreign, who
have so generously criticised a somewhat difficult
book to produce, the author's thanks are clearly due.
So fully and fairly have they done their work, that the
present volume has been undertaken with pleasure
and satisfaction, and their various suggestions have
been cordially adopted. An attempt has been made
to make the book in some sense worthy of the cause

it advocates and strives to advance, and in this spirit
the present edition is presented to the public. That
it may tend to improve the management, to increase
the popularity, and to extend the usefulness of
cottage hospitals in all parts of the world, is the
author's greatest hope and highest aim.

September. 1880.

EXTRACT FROM PREFACE TO THE FIRST EDITION.

IT came to my knowledge some months ago that there have been many and an increasing number of inquiries for a book on cottage hospital management. I communicated the fact to Dr. Swete, and asked him to issue a second edition of his work on the subject, but he informed me he was unable to undertake such a labour. Under these circumstances, and at the request of many friends, I venture to offer this little book for the guidance of those interested in the subject, under the belief that the present is a peculiarly suitable opportunity for the publication of such a work.

I have endeavoured to make myself familiar with the working and present management of the majority, at any rate, of the cottage hospitals which are open at the present time, and with this object I have visited many of them. For nearly ten years I have been a resident superintendent in a general hospital, either in the provinces or in London; and I have

availed myself of the experience thus gained in
hospital administration, in guiding me to a correct
conclusion as to the possibilities and requirements of
a model cottage hospital. I am highly indebted to
the medical officers and honorary secretaries of the
cottage hospitals throughout the country, and it is
to their cordial co-operation and assistance that I
owe the mass of information which is contained in
the tables at the end of the book.

March, 1877.

CONTENTS.

CHAPTER I.

ORIGIN AND GROWTH OF THE COTTAGE HOSPITAL SYSTEM.

PAGE

Introduction Early literature on the subject—Dangers of new buildings—The late Mr. Albert Napper, the founder of Cottage Hospitals—Counties without Cottage Hospitals—Number of Cottage Hospitals now in existence—Growth from Cottage to General Hospitals—Some failures and their causes—How to found a Cottage Hospital—Progress of movement—Causes of fluctuations in different years. - - - 1

CHAPTER II.

SOME MEDICAL AND SURGICAL ASPECTS OF COTTAGE HOSPITALS.

The former controversy as to comparatively successful treatment of cases in General and Cottage Hospitals—Mortality in Hospitals before and after antiseptics came into general use—Some difficulties experienced through lack of detail in Cottage Hospital reports—Suggestions as to, and forms to be used for giving details of patients treated—Midwifery cases in Cottage Hospitals—The late Dr. Matthews Duncan on the comparative mortality in large Lying-in Hospitals and in private practice—Midwifery cases if admitted to Cottage Hospitals to be treated in separate wards or buildings from other patients—The question of

(xv)

PAGE

midwives and midwifery in a state of transition—The
registration of midwives — Advantages of special
Cottage Hospitals for lying-in patients, -   -   - 19

CHAPTER III.

FINANCE.

Examination of present financial condition of Cottage Hospitals—The uniform system of accounts—Comparison
with expenditure in General Hospitals — Income --
Fallacy of arguments against Cottage Hospitals · Provident nature of such hospitals—Sources of income—
Annual subscriptions—Advantages over donations—
Large proportion of income derived from subscriptions
—Funded interest—Limitation of reserve fund—Donations—Collections in churches and chapels—Patients'
payments and their importance—Payments for medical attendance—Evils of indiscriminate free medical
relief—Payments for pauper patients—Objections to
ticket system—Good results of the free system—Form
of recommendation—Desirability of further extension
of system of patients' payments to General Hospitals, - 29

CHAPTER IV.

THE MEDICAL AND NURSING DEPARTMENTS.

Medical Department.—Advantages of Cottage Hospitals to
country practitioners and their patients — Mutual
advantages to consultants and general practitioners
—Economy of time and labour of country doctors—
Appointment of a medical director—Its advantages—
Duties of medical director—Professional intercourse
brought about by Cottage Hospitals—Cottage Hospitals as feeders to General Hospitals — Rules
applicable to medical staff,   -   -   -   -   - 59

Nursing Department.—Systems of nursing in force—Maintenance of regular supply of competent nurses—Cottage Hospitals must train their own nurses—Course

PAGE

of training for paying probationers—Advantages of
training paying probationers—Annual reports should
contain more details of the nursing department—
Ladies' Committee—Rules for the Ladies' Committee
of the Harrow Cottage Hospital—Advice and suggestions to nurses,   -   -   -   -   -   -   - 68

CHAPTER V.

DOMESTIC SUPERVISION AND GENERAL MANAGEMENT.

Ventilation—Necessity of constant interchange of air—
Daily inspection by head nurse—Flowers—Regularity
in administration of medicines and of food—Diet—
Family prayers — Bathing patients — Bed-making—
Draw sheets—Patients' clothes—Uneaten food—Bed
sores and their treatment—Administration of cod-liver
or castor oil—Preservation of ice—Administration of
alcohol — Treatment of accidents on admission —
Treatment for a sprained ankle—Enemas—Poultices
—Hot fomentations—Administration of food or medicine to semi-conscious patients—Fainting—Fits of
different kinds—Frost-bite—Foreign bodies in ear,
etc.—Dressing burns—Hæmorrhage—Injuries to head
—Operation room—Sponges.   -   -   -   -   - 80

CHAPTER VI.

SPECIAL FEATURES IN THE WORKING OF COTTAGE HOSPITALS.

Visits of patients' friends—Religious services—Office of
chaplain—Management—Letters of recommendation
—Medical practitioners not belonging to hospital
staff—Inadmissible cases—Patients boarded by the
nurse or matron—Practice at Fowey—Furnishing and
fitting-up a Cottage Hospital—Out-patients— Consultation by medical staff of hospital on cases sent by
medical men for that purpose—Hospital kitchen—

PAGE

Provident and other dispensaries Cleansing and
disinfecting at a given time—Varnished paper for
walls of hospitals and nurseries. - - - - 106

CHAPTER VII.

SPECIAL FORMS WHICH THE COTTAGE HOSPITAL MOVE-
MENT HAS TAKEN.

Cottage Fever Hospitals— Cottage Hospitals and paying
patients— Payments by patients according to their
means — Arguments for and against admission to
Cottage Hospitals of remunerative paying patients
Sir John Simon on pay wards at St. Thomas's
Hospital — Experience of Cottage Hospitals as to
mixing of free and pay patients—Special facilities
afforded by Cottage Hospitals for admission of remunerative paying patients—Opinions of managers
and medical officers of Cottage Hospitals on the subject
Convalescent cottages—Wards at Cottage Hospitals
for convalescent patients—Necessity for convalescent
institutions being under a separate roof from hospitals
--Convalescent homes needed by General Hospitals
Convalescent cottages for the poor of large towns —
The institution at Epping — Country holidays for
poor London children—Description of the efforts of
Madame Batthyany and Miss Synnot in this direction
—Mortuaries at Cottage Hospitals, - - - - 120

CHAPTER VIII.

COTTAGE HOSPITAL CONSTRUCTION AND SANITARY
ARRANGEMENTS.

Size—Considerations to be had in view—Site—Construction—Number of wards—Height—Plan—Preparation
of walls, etc. — Day-room for convalescents — Operating-room — Mortuary — Laundry -- Ventilation and
warming - Water supply -- Disposal of excreta—
Earth- and water-closets -- Drainage — Disposal of
sewage—Admission of enteric fever cases into wards
- Urinals—Disposal of slops and kitchen refuse. - 148

CHAPTER IX.

A MORE DETAILED ACCOUNT OF CERTAIN COTTAGE HOSPITALS, WITH PLANS AND ELEVATIONS.

PAGE

Description of the Cottage Hospitals at Cranleigh, Bourton-on-the-Water, Harrow, Petersfield, Harrogate, Lynton, Ditchingham, Leek, Walker, Ashburton and Buckfastleigh, Redruth, Petworth, Wirksworth, Reigate, Boston, Sherborne. Braintree and Bocking Chronic Hospitals—Rules at Saffron Walden. · 206

CHAPTER X.

SELECTED AND MODEL PLANS CRITICISED AND COMPARED, WITH A DETAILED DESCRIPTION OF VARIOUS HOSPITALS.

Description and criticism of plans of Cottage Hospitals at Grantham, Maidenhead, Petersfield, Ashford, Stamford (Fever), Bourton-on-the-Water, Beccles, High Wycombe, Cheshunt, Milford (Staffs), Forres Leanchoil, Watford, Mirfield, Falmouth, Brixham, Dartford, Wood Green and Surbiton—Plans for Model Hospitals of various kinds—Small Hospital with nine or twelve beds—General Hospital with thirty or forty-eight beds—Convalescent institutions—Temporary and permanent Fever Hospitals—Isolation Hospitals, Local Government Board's model plans—The construction of small Cottage Hospitals. - - 242

CHAPTER XI.

COTTAGE HOSPITAL APPLIANCES AND FITTINGS.

Surgical instruments—Dispensary requisites—Hair mattresses — Iron bedsteads — Linen and blankets — Counterpanes—Pictures in wards—Walls of wards—Lint and medical sundries—Patients' necessaries—Lockers—Brackets—Headings for beds—Ward furniture—Movable closets and baths—Hot-water plates —Screens—Bed rests—Arm slings—Surgical hammocks—Easy chairs—Foot rests—Filters—Boo

                                                              PAGE
shelves—Ambulances—Miscellaneous articles—Bags
and pads— Hot-water tins— Feeding cups—- Bed
pans. -   -   -   -   -   -   -   -   -   - 299

CHAPTER XII.

COTTAGE HOSPITALS IN THE UNITED STATES OF AMERICA.

Commencement and growth of the movement in the New
    England States- Lack of accommodation elsewhere
    in the United States—Baneful influence of politics on
    management of municipal hospitals--Superiority of
    those administered on the voluntary principle- Evils
    of indiscriminate medical relief—The constitution of
    Cottage Hospitals in the United States — Their
    financial system—Patients' payments--Municipal
    grants- Endowment of beds—Medical and nursing
    arrangements—Training of nurses in Cottage Hospi-
    tals—The matron's position—Register for nurses at
    Cottage Hospitals—Isolation of contagious diseases
    --Maternity cottages — Convalescent cottages —
    Hospital cottages for children — Cottage Hospital
    construction—Pay wards and their advantage in
    training of nurses,   -   -   .   -   -   - 327

Points upon which the Managers of Cottage Hospitals
    are requested to give special information in any
    communications they may address to the Author,   - 355

Model Forms of Annual Statements of Accounts, suggested
    for general adoption. -   -   -   -   -   - 356

Specimen page of Mr. Thomas Moore's Case Book, as
    used at Petersfield Cottage Hospital, -   -   .   - 358

Alphabetical list of Cottage Hospitals in (a) the United
    Kingdom and (b) the United States of America. -   - 359

A Set of Rules and Regulations for a Cottage Hospital,   - 365

ILLUSTRATIONS, PLANS, ENGRAVINGS, ETC.

PAGE

1. Full-page Engraving of the late Albert Napper, Esq., F.R.C.S., the Founder of Village Hospitals, with a view of Cranleigh Cottage Hospital . . *Frontispiece*
2. Diagram showing aspect of Hospital Wards, . . . 154
3. McKinnell's Double Tube Ventilator, ... 167
4. Sir Douglas Galton's Vertical Stove ; Plan, Elevation, and Section, ... 169
5. Thermoson Grate, ... 171
6. George's Coal-burning Calorigen, . . . 173
7. George's Gas do. do., ... 174
8. Slow Combustion Calorigen ; Vertical Section, Elevation and Plan, ... 175
9. Section and Plan of Ventilated Man-hole for disconnecting House-drains from Sewers. 191
10. Denton and Field's Automatic Sewage Meter. . 197
11. Sub-irrigation Drains, ... 198
12. Cranleigh Village Hospital, Elevation. . . *Frontispiece*
13. Do. do., Plan, ... 208
14. Elevation of Wirksworth Cottage Hospital. . . . 226
15. Reigate Cottage Hospital, Ground Plan, ... 227
16. Boston Hospital, Ground and First Floor Plans (*to face*). . 231
17. Grantham Hospital, Ground and First Floor Plans (*to face*), 243
18. Maidenhead Cottage Hospital, Elevation, ... 247
19. Do. do., Ground Plan, . . . 248
20. Petersfield Cottage Hospital, Ground and Chamber Plans, 250
21. Do. do., Elevation (*to face*), . 250
22. Ashford Cottage Hospital, Elevation (*to face*). . . 251
23. Do. do., Ground Plan, ... 251
24. Beccles Cottage Hospital, Ground and First Floor Plans (*to face*), ... 258

(xxi)

PAGE

25. High Wycombe Cottage Hospital, Ground Plan,   .   .   259
26. Cheshunt Cottage Hospital, Ground and First Floor Plans,   260
27. Sister Dora Convalescent Hospital, Ground and First Floor
    Plans (to face), ...   261
28. Forres Leanchoil Hospital, Ground Plan,   .   .   262
29. Watford District Cottage Hospital Plans (to face,   264
30. Falmouth Cottage Hospital, Ground Floor Plan,   .   .   270
31. Brixham Cottage Hospital, Ground and First Floor (to face).   271
32. Livingstone Cottage Hospital, Dartford, Ground Plan (to face).   272
33.   Do.   do.   do.,   First Floor Plan,   .   274
34. Wood Green Cottage Hospital, Ground Floor Plan,   .   275
35.   Do.   do.,   First Floor Plan,   .   276
36. Surbiton Cottage Hospital, Ground and First Floor Plans,
    (to face), ...   276
37.   Do.   do.,   Block Plan,   .   .   .   277
38. Model Pavilion Fever, General or Convalescent Cottage
    Hospital, with Mortuary, etc., Elevation,   .   .   278
39.   Do.   do.   Ground Floor Plan,   .   279
40. Local Government Board Model Isolation Hospital, for
    4 beds (to face), ...   286
41.   Do.   do.   do.,   for 8 beds
    (to face),   .   .   288
42.   Do.   do.   do.,   for 12 beds
    (to face), ...   289
43. Model Hospital Straight type, Elevation,   .   .   292
44. Do.   do.   do.,   Infectious Hospital,   .   .   293
45. Do.   do.   do.,   Ground and First Floor Plans,   294
46. Do.   do,   do.,   Villa Hospital,   .   297
47. Sketch of Hospital Bed-rest, ...   315
48. The late Mr. Sampson Gamgee's Surgical Hammock,   318

LIST OF STATISTICAL TABLES, MODEL FORMS, ETC.

PAGE

1. Table showing number of Cottage Hospitals established each year from 1855 to 1895 (July) . . 17
2. Form for stating number of beds in a Hospital 22
3. Form showing number of patients each year . . . 23
4. Form showing particulars of treatment . . . 24
5. Table comparing cost per bed in Cottage Hospitals, and in certain Metropolitan and Provincial General Hospitals 32
6. Table showing daily average number of beds, sources of income in 1892, and percentage of each source to total income of certain General Hospitals, 174 Cottage Hospitals, and 112 General Hospitals 35
7. Form of Letter of Recommendation 55
8. Form of Certificate of training for a Nurse at a Cottage Hospital 72
9. Diet Tables 84, 222
10. Table showing cost of alcohol at 18 Metropolitan General and 21 Cottage Hospitals in 1893 95
11. Form of Certificate for Country Home for London Children 141
12. List of Instruments necessary for Cottage Hospitals . . 301
13. Allowance of Linen per bed 305
14. Model form of Income and Expenditure Account . . 356
15. Page from Mr. Thomas Moore's Case Book at Petersfield Cottage Hospital 358
16. Model Rules for Cottage Hospitals 365

COTTAGE HOSPITALS.

CHAPTER I.

ORIGIN AND GROWTH OF THE COTTAGE HOSPITAL SYSTEM.

Introduction—Early literature on the subject—Dangers of new
buildings—The late Mr. Albert Napper, the founder of
Cottage Hospitals—Counties without Cottage Hospitals—
Number of Cottage Hospitals now in existence—Growth
from Cottage to General Hospitals—Some failures and
their causes—How to found a Cottage Hospital—Progress
of movement—Causes of fluctuations in different years.

WHEN the first edition of this book was published
in March, 1877, there were still many hostile critics
of the cottage hospital movement to be met with in
this country. Eighteen years' further experience has
proved to the medical profession and to the public
that the cottage hospital is an institution essential
to the well-being and comfort of all classes in rural
districts, and wherever the population is relatively
small and scattered. A few weeks ago it was our
good fortune to have the opportunity of visiting the
Savernake Cottage Hospital, which is beautifully
situated on the edge of the forest, and contains some
of the best small wards to be found anywhere. This

I

institution has enabled some of the medical practitioners in the district to acquire a skill in the treatment of surgical cases which is equal to that of the
leading surgeons of the day. There is no operation,
however difficult or complicated, which the staff have
not proved their ability to successfully perform in this
hospital. In those very difficult cases of injuries to
the head and face, the manipulative skill displayed
has been marvellous, and would do credit to the most
eminent surgeon. In such circumstances it is not surprising that the cottage hospital, when well managed,
as it usually is, enjoys a popularity equal to that of
any other class of hospitals wherever situated.

The early literature of cottage hospitals is very
interesting from the fact that it shows how precisely
those who first advocated the extension of this movement understood what they wanted, and how far it
would be possible to develop the system so that it
should supply the real needs of those whom it was
desired to succour without introducing abuses which
might undermine the character of the poorer residents
in country districts. The late Mr. Albert Napper of
Cranleigh was the author of the first work on the
subject; and his pamphlet,[1] which went through more
than one edition, and is now difficult to obtain, will
well repay perusal. His example was followed in 1866
by two other general practitioners, Dr. Wynter[2] and

---

[1] *On the Advantages Derivable by the Medical Profession and
the Public from Village Hospitals*, 1864; 3rd ed., 1866.

[2] Dr. Wynter, *Good Words*, 1st May, 1866.

Mr. F. H. Harris.[1] In 1867 Dr. Waring published a brochure;[2] and in 1870 Dr. Swete issued a small book[3] which was widely circulated and is still obtainable. The movement soon attracted the influential support of *The Builder* newspaper, which fully entered into the reasons for the establishment of these hospitals, and accurately estimated the success which was likely to attend the spread of the movement in rural districts. It is but bare justice to the early promoters and authors that we should pay a tribute to the wisdom and foresight they displayed in propounding a system for general adoption which has proved so successful in practice. Its simplicity, its ready adaptability to the requirements of country districts, and its successful working everywhere, have raised the cottage hospital system high in public esteem.

Here we would point a moral which is apt to be lost sight of in the present day by the more zealous advocates of the cottage hospital movement. Originally, the main principle advocated was simplicity in buildings, and indeed in everything. The present age has created a rage for new buildings; and it is within our experience, when paying visits of inspection to cottage hospitals up and down the

[1] *Remarks on the Establishment of Cottage Hospitals.* By F. H. Harris, M.R.C.S.; 1866.

[2] *Cottage Hospitals: their Objects, Advantages and Management.* By Edward J. Waring, M.D.; 1867.

[3] *Handy Book of Cottage Hospitals.* By Horace Swete, M.D.; 1870.

country, to find, that so long as simplicity prevailed and an old building was utilised for the purpose of the cottage hospital, the hygienic conditions of the building and the surgical results were good. In process of time, when funds became fairly abundant and were utilised to erect entirely new buildings and " modern " sanitary fittings and methods of drainage, their unhealthiness began to make itself felt, and the surgical results were attended with greater difficulties, if they were not even occasionally defeated, by the impurities which found their way into the wards owing to the adoption of a " modern " system of drainage.

We would therefore urge those who contemplate opening a cottage hospital in any part of the country to pay a visit to the Cranleigh Village Hospital, the original cottage hospital, and to the Wirksworth Cottage Hospital, which was established only a few years later than Cranleigh, that they may realise what excellent results are obtainable by the simplest means where the medical attendant is skilful and the old buildings are kept hygienically sound by the absence of all "modern" sanitary fittings of any kind. The truth is that, where there is no complete system of drainage in the village or township where the cottage hospital is situated, it is better for the health of the patients to avoid the " water carriage" system altogether, and to rest content with middens and earth closets situated outside and apart from the cottage or other building which may be converted to hospital use.

The late Mr. Albert Napper was undoubtedly the founder of cottage hospitals. It is true that the credit has been claimed for one of the sisters of the Holy Rood, who started a hospital with 28 beds in North Ormesby, Middlesbrough, in 1859. But the Middlesbrough institution was in no sense a cottage hospital; it was never heard of in connection with the cottage hospital movement; it was established in close contiguity to a large urban population, was too large originally to embody the real idea of a cottage hospital, and has now 60 beds. We may therefore dismiss this claim as untenable.

In 1859 Mr. Albert Napper organised the Cranleigh Village Hospital for the accommodation of the poor when suffering from sickness or from accident. What Mr. Napper meant by the poor is clearly deducible from the fourth of the rules for the government of the hospital, which provides that patients shall be received on the payment of a weekly sum, the amount of which, dependent on their circumstances, is to be fixed by their employer, in conjunction with the manager of the hospital. The admission of patients was granted by the manager on consultation with the medical officer. In his statement, published in 1864, of the advantages derivable by the public and the medical profession from the establishment of village hospitals, Mr. Albert Napper gave an account of the first hundred patients treated. Of these, 67 were parish paupers, 7 were incapable of remunerating the surgeon, 16 were in humble circumstances, and for

the remaining ten cases fees were paid by the poor
law guardians for operations, fractures, etc., to the
extent of £36 in all.

It is important to state these facts, because they
bear eloquent testimony to the character of the man
and the value of the work he did as the founder of
cottage hospitals. It will be seen that the scheme
which he originated has provided against the evils of
indiscriminate medical relief; has secured justice to
the medical profession, as every member is allowed
to follow his patient into the cottage hospital wards;
and, for the first time in England, has put the hospital
patient in the position of having the privilege of being
able to pay something, however small, according to
his means, for the treatment he receives. In consequence of the wisdom Mr. Napper thus displayed,
although there are at the present time some 300
cottage hospitals scattered up and down the country,
in every one of them the patients have the privilege of
paying according to their means, and from 8 to 10
per cent. of the total income of these hospitals is
derived from patients' payments. Had the same
system been followed by the larger hospitals, we should
not hear the universal complaint so justly urged
against them in regard to the abuse of their funds by
the treatment of well-to-do patients. But the scheme
has effected more than this. It has created a revolu-
tion for good throughout the whole of rural England.
By means of the cottage hospital the permanent
services of good nurses have been secured for

agricultural districts ; for the sick poor it has provided, for the first time in their experience, adequate accommodation in their midst, a sufficient supply of food and its due regulation, constant and regular medical supervision, and the removal of the dangers attending a long and wearisome journey, often with imminent risk to life or limb, and with the prospect of facing further and greater perils in the big town hospital if the patient survive the journey. To the country medical practitioner the cottage hospital has been a boon for many reasons. It has raised his professional status; it has enabled him to treat, under the most favourable circumstances, serious surgical cases which, before its institution, had to be transferred to the nearest county hospital ; it has established good feeling and friendly professional intercourse between the medical men of each district, for all have the right to use the cottage hospital, and to retain any fee which the poor law authorities may pay for an operation performed within its walls. The rich have not been slow to recognise the value to themselves of this increased experience and consequent skill on the part of the local practitioner. The lessons learnt day by day in the cottage hospital become in time of need of real value in the ancestral hall. Thus, as one writer has well put it, " the peasant's misfortune " has been " the means of saving the life of the squire."

Last, but not least, the cottage hospital has supplied a new source of interest for the inhabitants of the villages, and has united all classes in caring for the sick under the most favourable circumstances. It follows that Mr. Napper's work, as the founder of cottage hospitals, has a permanent value ; for it has resulted in great reforms of many kinds, and has secured to the great body of general practitioners a means of improving their experience, conserving their knowledge, and increasing their popularity and reputation with the population to whom they minister. The movement has been fruitful for good, and its founder deserves a high place in the annals of the medical profession.

It is satisfactory to be able to state from what Mr. Albert Napper himself told the author on one occasion, that the testimonial which was presented to him as the founder of cottage hospitals when he retired from practice some years ago, gave him greater pleasure than anything which happened to him in his long and useful life.

In 1880, when the second edition of this book was published, there were only five counties in England which did not possess at least one cottage hospital. Since that time, these institutions have multiplied at the rate of about 10 or 15 per annum, and at the present day (1895), there are only three counties which have not at least one cottage hospital. These counties are Huntingdon, Monmouth, and Rutland ; and it is not difficult to account for the isolated position they occupy in regard to cottage hospitals. These three counties have collectively a population of less than 350,000. Rutland (22,000) has no hospital accommodation whatever; Huntingdon (50,000) has one hospital, containing thirty-six beds (*i.e.* about one bed for every 1400 inhabitants); and Monmouth (275,000) possesses two hospitals, containing together fifty-nine beds (*i.e.* one for every 4700 inhabitants). Of the other two counties mentioned in the second edition, it will be observed that the influence of the Leicester Infirmary has not been sufficient to prevent the establishment of a cottage hospital in Leicestershire, namely, that at Hinckley (six beds) which was established in 1891. A district nurse is attached to it, who pays on an average upwards of 4000 visits a year, and relieves some 250 separate cases. Cumberland, the other county, now possesses a cottage hospital at Keswick (the Mary Hewetson Cottage Hospital), established in 1892, which contains, we believe, ten beds, although frequent application has failed to produce a copy of the report, or any particulars concerning this institution.

In our view, it is to be regretted that the medical profession in the county of Rutland have not yet succeeded in securing the establishment of, at least, one small hospital; for without it they are in a worse position for keeping in touch with modern medical and surgical progress than probably any other practitioners in the kingdom. We shall hope to hear that steps are being taken to establish a cottage hospital at Oakham at any rate. It is undesirable, in the best interests of the community, that the county of Rutland should continue to remain without any hospital accommodation within its borders for its population of 22,000.

On the other hand, in some counties where the movement took root at once, cottage hospitals are spreading with a rapidity which bids fair to soon bring the hospital accommodation up to the standard laid down by the best authorities, namely, one bed for every 1000 inhabitants in rural districts. There is, indeed, good reason to declare, in spite of the lukewarmness above referred to, that the cottage hospital movement has taken its place as one of the permanent charitable agencies of this country.

The actual position and proportions of the existing cottage hospitals must now be considered. It is probable that there are at the present time something like 300 cottage hospitals in the United Kingdom. This number will be slightly reduced when we have deducted those institutions which were originally started on the Cranleigh model, but which have been so far enlarged as to place them far above the scope of Mr. Napper's scheme, and to render them in fact small general hospitals for accidents. To these must be added the few hospitals which for various reasons were discontinued as failures. No better example of the first class can be selected than the North Ormesby Cottage Hospital, Middlesbrough, to which we have already referred. Originally started in 1859 with some 28 beds, it has been gradually increased until it has now accommodation for 60 patients, the in-patients amounting in 1894 to 552. The patients are admitted upon the recommendation of subscribers and governors. It is gratifying to find that the requirements of the district in which a cottage hospital is commenced have only to become known in order to ensure that the requisite hospital accommodation will be forthcoming. We find that many cottage hospitals have been enlarged more than once, an instructive fact of which Walsall may be taken as a striking illustration.

Turning to the list of failures given in the first and second editions of this book, the causes which led to such a result are in some cases so interesting and instructive that we reproduce them. It is satisfactory to add that in recent years we have not come across a solitary instance where a cottage hospital has been established and subsequently discontinued.

East Rudham was closed " not from want of funds, but really because the poor thought that the medical man, whose services were gratuitous, derived some unknown benefit from the hospital."

Southam had been closed for some years in 1880. It seemed that the patients were not numerous enough to warrant its continuance. It had undergone many changes, was first established in 1818 as an eye and ear infirmary, and had some endowment.

East Grinstead.—This is a sad exception, and it will be well to allow the medical officer to speak from his own experience. We quote from his letter : " In the completion of its eleventh year, after having had

over 300 cases under my care, I closed the hospital in
October, 1874. I did so for the following reasons:
In this district there are many very wealthy resident
and landed proprietors, but scarcely any volunteered
to help me. I got what money I required for furnishing it by writing direct appeals to individuals, and
afterwards to meet the current expenditure. I dare
say I could have done so up to the present year, but
after a time I became weary of making these appeals
to people who seemed to consider that, by contributing
to the support of the hospital, they were conferring a
favour upon me. This was especially annoying, as I
was not only giving my daily professional services,
but was also the greatest pecuniary contributor. In
addition to this, there were frequent attempts on the
part of wealthy people to get their servants or dependants into the hospital so as to avoid paying towards
their support. At length, after having experienced
for some years the meanness of the wealthy, and too
often the ingratitude of the poor, I closed the hospital,
and by the sale of the furniture raised a sufficient sum
to pay off the outstanding debts without any further
appeal to the charitable feelings of my wealthy
neighbours. I allowed the furniture and fittings to
remain in my cottage for upwards of two months
after I closed it as a hospital; and then as no one
offered to do anything to resuscitate it, I sent for an
auctioneer. Some time after, a few persons, who had
occasionally contributed to the hospital, met together,
and passed a vote of censure on me—the founder and

chief supporter of the hospital—for having closed it without obtaining their consent." [Subsequently, in 1888, a second hospital (5 beds) was started, and is now in existence.]

King's Sutton.—This hospital was opened in 1866 with 6 beds. It had 46 patients in its first year, and seemed to flourish. A letter addressed to the manager has been returned, marked "No Cottage Hospital now at King's Sutton." After every effort on our part, it has been found impossible to obtain further particulars.

Wrington.—This hospital was opened in July, 1864. It had a very prosperous career for about five years, when it was closed for the following reasons : It adjoined the grounds of the secretary—a dissenter. A discussion arose as to the patients attending his chapel. The surgeon in attendance maintained they should be allowed to go to church. To gain his point, the secretary bought the cottage hospital, and it was soon after closed, on the ground that there was no medical officer to attend to the patients. This is the only case of anything approaching a religious difficulty.

Great Bookham, opened in 1866, closed in 1868 as a hospital, and used for other purposes, such as providing nurses, monthly nurse, wine, brandy, etc., for the sick poor, at the discretion of the medical officer. Dr. Swete says : " For fear this should again enlarge to a cottage hospital, it is expressly laid down by the rules, that it shall be previously shown by the said medical man that the patient cannot, from the nature of the case, receive the necessary medical attendance at his or her own house, or that the patient's restoration to health is imperilled or impeded by the circumstances with which he or she is surrounded." He adds—and we heartily agree with him in this,— "Although numerous intended cottage hospitals have been strangled at birth, by opposition, this is the only case I have found in which a hospital once started has been put down. The animus against it must indeed be strong when the rules are framed so as to prevent the institution, which has risen from the ashes of the defunct hospital, ever again reverting to its original use."

Yate hospital was, unfortunately, closed, "though from no cause suggestive of failure of similar schemes."

It will be seen from the above instances that misunderstanding has in the main been the cause of failure. When it is considered that the failures do not amount to 2 per cent. of the whole number of institutions founded and opened, it will surely be admitted that the movement has been a great success. Wherever there has been a cottage hospital founded upon a broad unsectarian basis, with the requisite number of properly appointed officers, both lay and medical, to carry out the work thoroughly, there a success has been recorded.

It will perhaps be well to give here a few hints for the guidance of those who may wish to found a cottage hospital. These suggestions will prove of service to the charitably disposed, who, having experienced the want of such an institution, may be anxious to open one in their immediate neighbourhood. We are convinced that the first step to take is to call a public meeting, and to invite the attendance of every one in the district who is likely to take an interest in the movement, whether they may be favourable to the scheme or not. In a word, let the meeting be *public*, not only in name, but in fact. A little opposition at first is a positive advantage in the end, as it causes a wide spread interest to be excited, and the cottage hospital will thus become generally known.

If possible, the squire, or the rector of the parish in which the hospital will be situated, should be induced to take the chair. Two resolutions will be quite sufficient for all practical purposes. These should in effect be—(*a*) " That it is desirable that a cottage hospital should be started in this district," and (*b*) " That some five gentlemen, with the rector and the medical officer of the parish, be appointed a committee to carry out the necessary details." If this step be taken at the outset, it will disarm all suspicion of interested or personal motives on the part of any one, and, with ordinary tact in the conduct of the meeting by its promoters, its happy conclusion may be hailed with certainty as the commencement of a successful cottage hospital. Much responsibility will devolve upon the medical officer, to whose tact and good judgment the professional details must be left. Upon the exercise of these qualities the co-operation or hostility of the neighbouring practitioners will mainly depend. We shall treat of the medical administration of these hospitals elsewhere, and we merely refer to it here as one point of importance not to be overlooked at the outset. On page 365 will be found a model set of rules which will prove of assistance in drafting the rules for the new hospital. They are drawn up after a consideration not only of what is requisite to good management, but what it is necessary to avoid. They are submitted with confidence in their success, because they are the outcome of much experience in the working of these institutions.

We now come to consider the gradual increase of the movement, to mark its progress, and to ascertain if any and what circumstances have caused an increased interest in its prosperity.

It appears that of the whole number of cottage hospitals extant, so far as we have been able to ascertain, only eighteen were opened during the ten years immediately following that (1855) in which Mr. Napper commenced the movement at Cranleigh. So that, if we say the average number of new hospitals opened from 1855 to 1865 did not exceed two per annum, we shall be well within the facts. In the year 1866, however, no fewer than fourteen were opened, and the interest shown went on increasing steadily until 1870, when it reached its climax, and seventeen new hospitals were started. The average annual number established during the twenty-nine years

1866-1894, both inclusive, is about ten, the actual numbers being, we believe,--

1855-1865	.	.	.	18	Brought forward,		180
1866	.	.	.	14	1881	. . .	4
1867	.	.	.	14	1882	. . .	5
1868	.	.	.	14	1883	. . .	5
1869	.	.	.	14	1884	. . .	9
1870	.	.	.	17	1885	. . .	6
1871	.	.	.	17	1886	. . . .	7
1872	.	.	.	13	1887	. . .	9
1873	.	.	.	14	1888	. .	14
1874	.	.	.	6	1889	. . .	4
1875	.	.	.	7	1890	. .	6
1876				9	1891	. .	5
1877	.	.	.	5	1892	. .	9
1878	.	.	.	4	1893	. ,	9
1879	.	.	.	7	1894	. .	13
1880	.	.	.	7	1895 (to July) .		9

Carried forward, 180   Total, 294

What was it that acted as a stimulus, and caused so great a difference in the number established in one year as compared with another? Take, for instance, 4 established in 1865, and 14 in 1866; or the gradual decrease from 17 in 1871 to 4 in 1878. We have shown in previous editions of this book that the more public attention was called to the cottage hospital movement, the greater was the number of new cottage hospitals opened. The publication of the early pamphlets and the first and second editions of this book has had a material influence upon the movement, and has led to increased activity in the establishment of cottage hospitals. It will be seen that during the last twenty-one years there were more cottage hospitals opened in 1888 than in any other year. This is

2

accounted for by the circumstance that in many places the inhabitants determined to commemorate the Queen's Jubilee by the erection of a cottage hospital. We venture to think that it was impossible to hit upon a better memorial of the beneficent reign of H.M. Queen Victoria than the establishment of one of these institutions in a district which has heretofore long felt the need of adequate hospital accommodation. We hope that the issue of the present edition of *Cottage Hospitals* will stimulate other communities to consider the advantages which would result from the opening of a small hospital in their midst. It has been a source of deep regret to the author that owing to demands upon his time it has not been found possible to issue a third edition until the present year. He has always been anxious to do his utmost to encourage the growth of cottage hospitals ; and the interest which is still taken in this movement is shown by the circumstance that the publishers have had an ever-increasing number of orders for copies of this book which they have been unable to supply because the previous editions have for years been out of print. The author has, however, done his utmost to afford information or assistance to those who were establishing cottage hospitals ; and he hopes that the present volume may once more extend the interest in these useful institutions, so that the day may not be far distant when a cottage hospital will be found wherever its establishment is calculated to promote the well-being of all classes of the community.

CHAPTER II.

SOME MEDICAL AND SURGICAL ASPECTS OF COTTAGE HOSPITALS.

The former controversy as to comparatively successful treatment of cases in General and Cottage Hospitals—Mortality in Hospitals before and after antiseptics came into general use—Some difficulties experienced through lack of detail in Cottage Hospital Reports—Suggestions as to, and forms to be used for, giving details of patients treated—Midwifery cases in Cottage Hospitals—The late Dr. Matthews Duncan on the comparative mortality in large lying-in hospitals and in private practice—Midwifery cases if admitted to Cottage Hospitals to be treated in separate wards or buildings from the other patients—The question of midwives and midwifery in a state of transition—The registration of midwives—Advantages of special Cottage Hospitals for lying-in patients.

Up to 1880 much controversy existed as to the comparative successful treatment of cases in large general hospitals, notably in the metropolis, and in cottage hospitals. In the second chapter of the last edition of this book we published a table by the late Dr. J. C. Steele, of Guy's Hospital, showing the mortality at all the London hospitals of importance, general and special, during the years 1872, 1873 and 1874, and tables

(19)

showing the death-rate in 40 cottage hospitals in 1874 and 1875. These tables showed that the results attained were much better in cottage hospitals than in the great metropolitan general hospitals. Since that day, owing to the strict enforcement of the antiseptic system and to improved hygienic conditions, the statistics obtainable for general hospitals in London are far more favourable at the present time than they were fifteen years ago. So much is this the case that the most recent statistics prove that there is little or no point to be made in favour of the cottage hospital; for under present conditions septic disease has become so comparatively rare in ordinary cases as to place all hospitals, whether large or small, upon practically the same footing so far as results are concerned. Schede's statistics illustrate the extraordinary value of Sir Joseph Lister's methods, and it may be interesting for this reason to quote them here. Thus, of 387 cases treated in the pre-antiseptic era 29.18 per cent. died, while out of 321 cases treated during the antiseptic era only 4.4 per cent. died. These figures show at a glance how much better the present results are than those of ten years ago. Dr. Frederick Page, the eminent Newcastle surgeon, has published the results of major amputations treated antiseptically in the Royal Infirmary, Newcastle-upon-Tyne, for a period of twelve years and nine months, ended 31st December, 1890. These statistics have been most carefully compiled, and have a special interest. They show that out of 606 cases 47 died—a mortality of 7.7 per cent. Of the whole number of cases 237 were operated upon for injury, with 28 deaths, equal to 11·8 per cent.; and 369 were operated upon for disease, of which 19 cases proved fatal, equal to a mortality of 5·1 per cent. The results recorded in cottage hospital practice, on the whole, continue to be excellent; and as the controversy is now practically at an end, we have determined to omit the tables given in former editions, for although they have an historical value the altered conditions of surgical practice everywhere in the present day have practically settled the question.

One important point remains, however, and we commend it to the candid consideration of all medical officers attached to cottage hospitals. Not infrequently we observe a serious want of exactness in stating the particular diseases for which the patients were treated. It may be mentioned, in the first place, that it has been found impossible to give the ages of all the patients, as these particulars are usually omitted from the tables of cases given in the cottage hospital reports. In many instances the sex of the patient is not even stated, whilst the cause of death is omitted altogether. Of late years much more attention has been paid to these details by cottage hospital authorities, and we look forward to the day when any such difficulties will disappear. Of course, the omission to state the cause of death may be due to the difficulty of getting a *post-mortem* examination in such cases, but we cannot too strongly impress upon each medical man that the immediate cause of death should always be clearly ascertained and recorded.

We ventured many years ago to formulate a plan, including the form of a table, in which returns might be made of the diseases treated each year. If this were universally adopted in cottage hospitals (as already has been done by the best managed of these institutions), and if the information were published in their reports, those reports would have a permanent value which they too often at present lack. We would further point out that it is the exception to find the number of beds in a hospital stated. This information should always be given, and might be printed upon the cover of the annual report. Thus :—

" ....COTTAGE HOSPITAL.

" No. of beds, 8. Average number occupied during year 1894, 6."

While on the subject of tables and statistics, it may be well to impress upon the staff of cottage hospitals the advisability of keeping an accurate account of each case treated, and of carefully recording in a medical case-book specially kept for that purpose the condition of the patient when admitted, the progress made, the treatment adopted, and the result. In addition to this, it will be necessary to publish annually a list of cases, with full particulars of the patients' ages, sex, dates of admission and discharge, the injury or disease, and result, giving in the case of

a death the immediate cause, and the number of days
the patient was in hospital.   For the sake of simplicity, and as a guide to those who may desire to
follow this plan, we append a form (page 24) which
we recommend for general adoption.   It would be
an additional advantage to give a short summary
of all the cases under treatment, thus :—

*During the year   in-patients have been admitted (of whom
        remained under treatment on 31st December, 189), with
    the following results, viz. :—*

Cured,
Relieved,
Incurable, .   .
Died, ...
Remaining in hospital,   .

                              _____ ___

                              Total

*The average cost of each bed has been £   ; the average
cost of each patient was £   ; and the average number of
days each patient remained under treatment was   . The total
available beds have been   ; and the average number
occupied throughout the year was   .*

List of In-Patients treated in the   *Cottage Hospital,*

from   189 *to*   189 .

Number of Beds   Average Number occupied through the year

No.	Initials.	Age.	Sex.	Occupation.	Days in Hospital	Diagnosis. State when admitted.	Treatment.	Result.	REMARKS. Short History of Cases. (Plenty of space should be allowed for this column.)

In cases of death after an operation, state whether the patient died from secondary hæmorrhage, shock, pyæmia, or other cause, and give a short history of all such cases.

Number of Out-Patients treated from   189 to   189 .

In giving a list of the patients it is very desirable that a short account of each case should be given in the column headed "Remarks." This course has been carried out with excellent method by Dr. Beeby, at the Bromley Cottage Hospital. We extract three of his cases as examples of the kind of history we think it most desirable to give :—

"*M.* Aged six. Son of mill-hand, St. Mary's Cray. Admitted 5th July, suffering from injuries caused by carriage accident. Discharged on 24th August in a very improved condition. Was sent to the Royal Sea Bathing Infirmary, Eastbourne.

"*R. C. L.* Aged forty-five. Trowbridge. Admitted on 16th September, suffering from injuries received in fall from dog-cart. Concussion of brain—probable fracture of base of skull. Died on 17th September. Inquest—verdict, accidental death.

"*J. W. D.* Aged twenty-three. Labourer, Farnborough. Admitted on 26th September with severe compound fracture of humerus. Amputation. Discharged cured on 7th November."

A short history of each case, thus given in the annual report, causes much interest to be excited in the work of the cottage hospital, and proves alike interesting to the subscribers and beneficial to the institution. We hope to see this example widely followed.

In the last edition we devoted considerable space to the following two questions :—

(1) Should the large lying-in hospitals be done

away with, and small cottage hospitals built in their place? and (2) Is it advisable to treat midwifery cases in cottage hospitals rather than at the patients' own homes? We showed from the evidence published by the late Dr. Matthews Duncan, in his book on *Mortality in Childbed and Maternity Hospitals*, that the large hospitals are not necessarily more unhealthy than the small ones; that there is always a certain proportion of deaths to labours even in private practice; that, taking the mortality of the Rotunda Hospital, Dublin, as the largest, and the one with the most complete set of statistics, its mortality would challenge comparison with that of any small hospital, or with that of any private practitioner whose statistics could be confidently relied upon. On the whole Dr. Duncan may be taken as having proved that large hospitals, if well built and properly managed, are quite as successful as small ones; but the controversy cannot yet be considered settled, and it will probably be waged with varying success for many years to come.

A reply to the second question must depend largely upon the circumstances of the population in particular districts, and the means which the charitable public are prepared to place at the disposal of the local practitioners or hospital authorities. If, as often happens, poor women in a country district are delivered under miserable conditions by ignorant and prejudiced midwives, the advantage of maintaining a cottage hospital for lying-in cases cannot be questioned.

There is a great fallacy of assumption in taking for granted an excessive mortality in midwifery hospital practice; but there is, unfortunately, no fallacy as to the miserable conditions of many poor women in rural areas, where ignorance and prejudice still prevail to an extraordinary extent. If, then, funds are forthcoming, we should favour the erection and maintenance of a small cottage hospital for midwifery cases. The plan has not yet found sufficient favour to ensure a trial of the system, but indirectly it has already proved successful. Not a few cottage hospitals admit parturient women into their small wards, and the results have so far been encouraging, though we do not think the plan ought to be permitted unless a part of the building be set apart for these cases, with a separate nurse or attendant. We regard separate wards as absolutely essential; and at Milton Abbas and elsewhere, midwifery cases are admitted by permission of the committee, who, it is fair to presume, take every precaution to obviate all risks to the patients.

The whole question of midwives and midwifery is now in a state of transition. On the one hand, there is a strong feeling that midwives should be registered, to which we incline; and, on the other, an agitation against registration by a certain class of medical practitioners, who maintain that every woman should be delivered by a duly qualified medical practitioner,—a contention which can never be maintained if any due regard is to be paid to the require-

ments of the poor women and to the fees which every
self-respecting doctor ought to insist upon receiving
for these cases.

Our hope and belief is that some day the many
great advantages to be gained by taking midwifery
cases into cottage hospitals built for the special
purpose, must lead to their establishment. Such
institutions for lying-in cases would be a priceless
boon to many poor women, by removing them from
home troubles, and worries, and securing a more
strict regard to rest, cleanliness, and decency. It is,
we believe, impossible to do without midwives, who
attend the poorer classes at fees for which it would
not be possible for a hard-worked country doctor to
give up his time, and the Lying-in Cottage Hospital
would enable a better class of midwives to be trained
by the general practitioner, and at the same time
secure that the doctor should always be called upon in
cases of emergency. With regard to fees, too, the
cottage hospital is based on the principle that patients
ought to pay a certain sum, instead of being entirely
dependent upon charity. It should be easy therefore
to arrange a scale of fees, by which patients would
have all the advantages and comforts we have indicated
without unnecessarily crippling them, and at the same
time adequate remuneration would be secured to the
medical attendant for every case where his services
were essential to the well-being of the patient.

CHAPTER III.

FINANCE.

Examination of present financial condition of Cottage Hospitals—The uniform system of accounts—Comparison with expenditure in General Hospitals—Income—Fallacy of arguments against Cottage Hospitals—Provident nature of such hospitals—Sources of income—Annual subscriptions—Their advantages over donations—Large proportion of income derived from subscriptions—Funded interest—Limitation of reserve fund—Donations—Collections in churches and chapels—Patients' payments and their importance—Payments for medical attendance—Evils of indiscriminate free medical relief—Payments for pauper patients—Objections to ticket system—Good results of the free system—Form of recommendation—Desirability of further extension of system of patients' payments to General Hospitals.

IN judging the influence and position of a country, it is usual to obtain, if possible, an account of its revenues, together with its liabilities and financial prosperity ; and these may fairly be taken as an index of its right to rank as a first or as a second rate power. This is equally true of systems and of institutions. We therefore propose to examine the present financial condition of those cottage hospitals where the work done has been continuous and efficient.

During the last few years it has been found possible to induce the managers of all the best administered hospitals and kindred charities in this country to adopt the same uniform system of accounts. It is most desirable that cottage hospitals should follow a similar course. In the hope that this might result we have published a book on a uniform system of accounts[1] and have devoted a chapter to the consideration of how to apply the uniform system to cottage hospitals and small institutions. The plan recommended reduces the books to two in number, namely, a cash analysis, receipt and expenditure book, and a secretary's or matron's petty cash book.

The hon. secretaries or matrons of cottage hospitals and the managers of smaller institutions will find the two books here recommended amply sufficient for their requirements. By using the index of classification, they will be able to analyse and post each item under its proper heading, and so their accounts will agree in every particular with the accounts kept on the double-entry system by the larger institutions. In the annual report, secretaries who adopt the uniform system here recommended will publish the income and expenditure account for the year in the form given on page 356. We have included in this account a great number of items which will not be required by the smaller institutions, but they are published because every hospital has its own special features, and by giving the whole of the items it will be easier for the secretaries to select those which are wanted to enable them to set forth

[1] *The Uniform System of Accounts for Hospitals and Public Institutions* (London: The Scientific Press, Limited, 428 Strand, W.C.).

clearly and in detail the items which apply to their cottage hospital. Those who procure the two books recommended from The Scientific Press, Limited, will see that in the income and expenditure account which is given, the income side consists of the headings and totals of the columns in the cash analysis, receipt and expenditure book ; whilst the expenditure side, on the other hand, contains the headings and totals in the secretary's petty cash book. It will, of course, be understood that under this system all payments must be entered in the petty cash book, and that all accounts will be paid out of petty cash. We may add by way of explanation, that the individual figures in the first total column of the cash analysis, receipt and expenditure book should be made to correspond with the individual entries or totals on the receipts side of the bankers' pass book. Similarly, the individual items appearing in the second total column of the cash analysis, receipt and expenditure book should correspond with the counterfoils of the cheque book, and so represent the individual items or figures on the expenditure side of the bankers' pass book.

The average annual expenditure in 1892, in 183 cottage hospitals, having an average number of from twelve to fifteen beds, and in 106 of which the average number of beds occupied was eleven, was £622. This gives the cost per bed as £41 on the whole number, or £66 per bed occupied. When it is considered that the average cost per bed at a hospital like the London Hospital was in 1892 £79, or £97 per bed actually occupied, while at the Middlesex Hospital the cost was respectively £90 and £112, it will be seen that, on the score of economy, the cottage hospital has much in its favour. The cost in cottage hospitals compares also very favourably with the provincial general hospitals; for we find that at the General Hospital, Birmingham, each bed occupied cost, in 1892, £67, at Leeds £62, and at Edinburgh £64.

The hospitals selected in the following Table may be fairly considered as types of the different kinds of management now in force at the general hospitals in the country.

Table showing the relative cost per bed in 183 Cottage Hospitals and in certain Metropolitan and Provincial Hospitals, compared.

HOSPITALS.	Number of Beds.	Cost per Bed.			Actual Number of Beds occupied.	Cost per Bed occupied.		
		£	s.	d.		£	s.	d.
183 Cottage Hospitals	2792	40	15	2	1170[1]	66	9	1 [1]
London Hospital	776	78	11	9	630	97	0	0
Middlesex Hospital	307	90	4	10	254	111	14	4
Westminster Hospital	205	64	15	10	186	71	8	3
Newcastle Royal Infirmary	270	46	18	6	251	50	9	6
Leeds General Infirmary	436	44	13	5	315	61	10	4
Birmingham Gen. Hospital	271	57	11	0	233	66	18	9
Edinburgh Royal Infirmary	710	59	2	3	652	64	7	5
Glasgow Royal Infirmary	582	55	8	2	560	57	11	8
Belfast Royal Infirmary	189	34	2	4	141	45	14	7

In the above Table the extraordinary expenses (*i.e.*, cost of building and permanent improvements) have been deducted in calculating the costs per bed and per bed occupied.

When the difficulties which surround cottage hos-

[1] Only 106 hospitals give the number of beds occupied.

pitals are considered,—the necessity of buying in small quantities, and the consequent loss of trade discount, the cost of carriage, the difficulties with respect to water, drainage, and many other matters— the result, as shown by the figures, must be gratifying to the original promoters of these institutions. It is a great feather in the cap of cottage hospitals, that the expenditure per bed, during the last fifteen years, has not materially increased. Fifteen years ago the cost per bed occupied in a cottage hospital was £66 7s., and as the above figures show it is now only £66 10s.

Let us next turn to the income of these hospitals, and we shall find the result still more satisfactory. The average income, including legacies, of each of the 183 cottage hospitals above alluded to, is annually £720, in round numbers, or nearly £100 per annum more than the average expenditure. This seems to point to thrift and good business management. Thus, whereas the average annual expenditure at 183 cottage hospitals was £113,798, the average income was not less than £131,484, leaving an ample margin in each case for emergencies. Here, then, is unmistakable proof that in spite of hostile criticisms and innumerable difficulties at first, the cottage hospital has quietly but surely pushed its way onwards, until it is fairly entitled to hearty recognition at the hands of the profession and the public everywhere. The time has evidently gone by when it could be said :—Is it wise to encourage the growth of these little hospitals? Will they not interfere with the work of the great hospitals in our county and manufacturing towns? By doing the work at greater expense, but with less efficiency, will they not leave us eventually with little reliable or efficient hospital accommodation any-where? There were many people who did not scruple to say this and much more thirty years ago, and we shall like to see what they will urge now that their prophecies have ended in nothing. Was there ever a time when more anxious consideration was being given to the subject of hospital accommo-dation and hospital efficiency than the present? We have been surprised to find how many thousands of pounds have been spent throughout England in recent years on hospital extension. Has any single county infirmary or large general hospital suffered any loss of income by the cottage hospital movement? Few will maintain such to be the case. Yet some £200,000 a year has been raised, in addition to the sums spent on the large hospitals, by a new agency, but for somewhat similar purposes. This simple statement of facts is unassailable, except from one point of view,—the provident. We have no doubt many people will feel inclined to argue somewhat thus :—Surely, if the poor of this country got on fairly well in sickness before the origin of this movement, and if there was always room and to spare in the large general hospitals, you have tended by this additional relief to pauperise, rather than to benefit the masses.

We propose to meet this view with further facts in reference to the income of cottage hospitals, and

AVERAGE NUMBER OF BEDS AND SOURCES AND PERCENTAGES OF INCOME.

Table showing daily average number of Beds, sources of Income in 1892, and percentage of each source to Total Income of certain specified General Hospitals, 174 Cottage Hospitals, and 112 General Hospitals in London and the rest of England, Scotland and Ireland.

Hospitals.	Daily average number of Beds.	Subscriptions.	%	Donations.	%	Church Collections (including Hospital Sunday and Saturday).	%	Funded Interest.	%	Patients' Payments.	%	Miscellaneous.	%	Total.	Legacies.
		£		£		£		£		£		£		£	£
174 Cottage Hospitals	—	40,506	36	13,137	12	25,476	23	14,076	12	9,328	8	9,885	8	113,408	13,259
112 General Hospitals	—	215,029	23	185,666	20	110,930	12	347,867	38	19,049	2	39,747	5	919,288	294,745
Middlesex	254	2,881	16	3,630	20	2,371	12	9,154	49	—		470	3	18,506	31,035
Westminster	186	1,473	19½[1]	1,213	16	1,393	19	2,734	37	563	8	29	½	7,405	12,403
Newcastle Royal Infirmary	251	3,273	28	3,806	32	1,223	10½	3,003	26	23	¼	364	3	11,692	5,010
Leeds General Hospital	315	4,913	23	3,222	15	6,884	31½	5,999	28	352	1½	300	1	21,672	926
Birmingham General Hospital	233	5,355	32	226	1	7,550	45	3,408	21	—		151	1	16,690	10,697
Edinburgh Royal Infirmary	652	3,534	4	3,575	11	—		9,295	30	499	1	4,375	14	31,278	11,380
Glasgow Royal Infirmary	560	7,309	34	8,704	39	—		5,629	25	73	⅓	388	1⅔	22,163	6,613
Belfast Royal Infirmary	141	1,671	16	5,818	58	591	6	1,094	10	706	7	317	3	10,197	3,151

[1] Triennial Grant.

to show from them, that, as a system, this movement
has done more good in a provident sense than the
general hospitals have ever thought of attempting.
What are the sources from which the cottage hospitals
derive their income? They are chiefly five, namely,
annual subscriptions, donations, church collections,
(including Hospital Sunday fund), patients' payments,
and interest on funded property. The proportion
per cent. of each source of income in the foregoing
order is 36, 12, 23, 8, and 12. We think this fact
shows clearly that from the first this movement has
been guided and advanced with a regard for sound
management which is truly remarkable.

Every one versed in hospital management knows that
the secret of financial soundness in things charitable is
summed up in the successful attainment of a large proportion of the income from annual subscriptions. If
a man gives a donation, he gives it in a comparatively
thoughtless way, with the feeling that he has done a
good act, and, in doing his duty thus, has done all
that is necessary, so far as the particular institution
is concerned. He does not, as a rule, care to see a
report, or to hear anything more about it. He has
given a good round sum—the actual amount, whether
small or great, does not make a material difference in
this feeling—and he does not wish to be bothered
further in the matter. If you ask such an one, some
five years later, if he will attend a meeting of governors,
and vote for some particular person or thing in which
you happen to be interested, he will probably express

surprise that he is governor, and will candidly confess
that he had forgotten alike his donation and the
institution, until your request put him in mind of both.
On the other hand, if a person becomes an annual
subscriber, he hears of the charity at least once a
year, and probably, before paying his subscription, he
will turn to the annual report, which had doubtless
been thrown aside with many other similar docu-
ments, and will mentally decide, from what he reads,
whether he is satisfied with the management of the
charity or not. If not, he will in most cases, whilst
giving his subscription, either express his views to the
person who calls for it, with a request that they may
be brought to the notice of the managers, or he will
bear it in mind, until he has an opportunity of calling
at the institution itself.

Experience has shown that, as a rule, legacies are
given to charities by the annual subscribers rather than
by the life governors or donors. This is a striking fact,
and one well worthy the attention of all who have the
management of these institutions. It is much better
policy to try and obtain an annual subscription of *one*
guinea, than to accept a proffered donation of *ten*. The
donor, as a rule, patronises all charities alike, and you
find the same names occurring over and over again, for
the same sum in very many reports. The annual subscriber, on the other hand, gives a smaller sum to some
three or four charities, but he gives it after a due consideration of the respective merits of each individual claim,
and, having once given his adhesion to the management

of a charity, he not only pays his subscription with the
regularity of the dividends, but he reads the annual report, and *misses it* if he does not receive it at the usual
time.   He attends meetings of the governors at least
once in the year ; he recommends the charity to his
friends ; and, when he dies, the names of all the institutions he has thus loyally supported during his lifetime
will be remembered without fail in his will.   There is
nothing the class from which donors are drawn dread
so much as the thought of being " bothered," as they
call it.   That is to say, they decline to carefully weigh
the respective merits of particular charities.   Hence
it used to be no uncommon occurrence, in London at
any rate, for small charities to be started by venture-
some persons who had a knowledge of this fact.   It
is well known to many people behind the scenes, that
not only are such charities maintained by these
thoughtless donors, but that their originators not infrequently reap large benefits personally from the
contributions of the same misguided individuals.

Bearing these facts in mind,—facts which we hope
may be brought home to the charitably disposed as
well as to hospital managers,—it is very gratifying to
find that more than one-third the ordinary income
(exclusive of legacies) of the cottage hospitals is
derived from annual subscriptions, so that quite
£70,000 every year is raised by this means.   It is
probably not known that this sum exceeds the
amount of the subscription lists of all the London
general hospitals put together.   In fact, it represents

a sum equal to half as much again as the total income
from annual subscriptions of the whole of the general
hospitals in the metropolis which are supported by
voluntary contributions, although they have collectively upwards of 4100 beds to maintain. It has long
been argued, that it is impossible to get annual
subscriptions for London charities, because people
now-a-days positively refuse to pledge themselves to
give a fixed sum annually to any object.

Besides, the cottage hospital supporters are all
familiar with its object, its management, and its work.
They live close to it, or frequently pass it in their daily
rambles, and so it is easy to persuade them to show a
permanent interest in its welfare by becoming annual
subscribers. The answer to these objections is
simple. The experience of some of the London
hospitals, where special attention has been given to
this subject, proves unanswerably that it is quite
possible to get annual subscriptions if sufficient
labour be expended. No doubt the cottage hospital
is a centre of interest to those who live near it. But
without continuous and well-directed efforts it would
not have been possible to raise such an unusual sum
as the managers have done, and are still doing.
Besides, do not these facts show the importance of
getting annual subscribers to attach themselves to the
various charities, on account of the active interest
they will then take in their welfare? Surely it is
worth while to use every legitimate means for securing
annual subscriptions, and let all honour be paid to

the promoters of cottage hospitals for the wisdom and business acumen they display in raising the necessary income for their work.

Let us next take the funded interest as a source of income, and we shall find that this item, with legacies —which do not figure largely in cottage hospital accounts—make up more than half the whole income (ordinary and extraordinary) at the London general hospitals ; whereas something like 22 per cent. only of the total income of the 174 cottage hospitals analysed in the Table is derived from these sources. We shall reserve our remarks on the proper disposal of surplus income where the system of patients' payments is in force, in preference to investing it as an endowment fund, until we treat of the relation of the medical profession to these little hospitals. Meanwhile we may express our regret that so much gratuitous work is done by the medical staff, when 174 cottage hospitals have a total annual income from invested savings of £14,076. It is surprising to us that any person should ever give anything to a charity without first ascertaining its exact financial position, and still more surprising that ladies and gentlemen, when they find a charity has a large endowment fund, should consent nevertheless to give considerable donations towards its maintenance, not nominally indeed, but virtually, to enable the managers to accumulate a large surplus property, which will probably one day be diverted to some other object by the "Charity Commissioners" of another generation. Is it, or is it

not the fact, that where there are large invested funds
at the disposal of a charity, far in excess of the necessities of the case, there, sooner or later, bad management and reckless expenditure will probably find a
home? Long experience has taught us that the best
managed charities are those that are not overburdened
with invested capital. Of course there are exceptional
reasons, in isolated cases, where it is necessary for the
managers to secure a considerable reserve fund, to
meet the possible exigencies of their particular case.
It may, however, be calculated roughly, if a charity has
a reserve in the way of capital to the amount of two
years' expenditure, it has quite as much as it is wise
for the charitably disposed to allow it to possess.

There is something almost grotesque about a request for a donation, which follows the statement
"that by careful management we have been able
to invest £100,000," and yet this is said of many
of our noblest charities over and over again. We
earnestly hope that the managers of cottage hospitals will not attempt to accumulate any capital.
Let them purchase the freehold of their land and
buildings, and having done this, and having provided every necessary appliance and comfort for the
patients, let them consider the suggestions made
elsewhere as to the remuneration of the medical staff,
and let them act upon them. We look upon the fact
that in 33 years the 174 cottage hospitals in the table
on page 35 have secured a capital amongst them of
something like £350,000 with great apprehension for

the future economical management of these charities. This item has nearly trebled in the last fifteen years, a conclusive proof of the necessity of again cautioning the managers against so baneful a practice. If cottage hospitals are to maintain their character for inculcating provident habits amongst the labouring classes by means of small payments, the system of endowments must find no permanent home amongst them.

Donations are, of course, a very useful source of income, and are not by any means to be discouraged or despised. They must be regarded, however, as the means by which a certain elasticity of income can be secured, and not as a reliable basis for judging of prosperity or soundness. Under these circumstances, it is pleasing to find that cottage hospitals derive about 12 per cent. only (or about a tenth of their total income including legacies) from this source, whereas general hospitals in the United Kingdom depend upon donations for about 20 per cent. of their total ordinary income excluding legacies.

The next important item of income is that derived from collections in churches and chapels. It is greatly to the credit of all classes of the community, that a movement which had its origin some thirty-six years ago in the pressing necessities of the largest general hospital in Birmingham, should have gained so strong a hold upon the religious communities throughout the country. Not only in the large towns, from London downwards, but in many villages and hamlets, " Hospital Sunday " has grown to be recognised as a great annual festival for the relief of the sick. This system has been introduced into America, and the first Hospital Sunday in that country was instituted in New York fifteen years ago. Nearly one-quarter of the income (23 per cent.) has been procured by this means for the 174 cottage hospitals here taken. The actual annual sum thus raised is about £26,500. The early leaders of the movement urged upon the promoters of cottage hospitals the importance of securing the co-operation of the clergy in the management of these institutions ; but, we will venture to say, they never thought of the large sums of money these gentlemen would be the means of bringing into the hospital coffers. Indeed, when Cranleigh Hospital was opened in 1859, the first Hospital Sunday collection had not even been made at Birmingham.

We have no space, however, to fully deal with the Hospital Sunday movement here ; but those who are interested will find all the facts succinctly stated in *Burdett's Hospital and Charities Annual* (London : The Scientific Press, Limited, 428 Strand, W.C.), 1895, and previous editions.

We have only one other source of ordinary income to refer to, and our list is exhausted. We allude to patients' payments.[1] The actual sum derived from this source by the 174 cottage hospitals on our list is about £9500, or about 8 per cent. of the total income, whilst at the 112 general hospitals it only provides about 2 per cent. of the ordinary income.

[1] See also Chapter VII.

Patients' payments are a special feature of the cottage hospital system, and they were practically unknown to the managers of general hospitals in this country until fifteen years ago. Cottage hospital managers from the first have recognised the importance of encouraging feelings of self-help and independence amongst their patients. Hence, without exception, the system of patients' payments is to be found in force at all true cottage hospitals. We use the word "true" because in a few instances some have tried to abolish the system. In future no hospital so managed should be regarded as a cottage hospital, and for this reason.

The original promoters of this system of relief desired to help the poor and not to pauperise them. Any institution therefore which fails to carry out the plan of patients' payments forfeits its right to be considered a cottage hospital. It may be a private charity of a few wealthy personages, or an almshouse, or a sort of private poorhouse; but as its principles of management are diametrically opposed to those which have made these small hospitals thrifty and popular, it certainly cannot claim the proud title of "Cottage Hospital." If ever these small hospitals become free to any large extent, owing to the trouble which is caused by receiving or collecting the patients' payments, or because the committee have a considerable endowment, from that day they will prove a curse rather than a blessing to the labouring poor. This is one of the rocks ahead. Those, then, who really have the welfare of the cottage hospital at heart, will take care that the love of patronage which has so strong a hold upon some otherwise worthy people, shall not be allowed to ruin so beneficent a system of medical relief as that of patients' payments, which originated with the cottage hospital, and which forms the brightest jewel in its crown. The sums paid are very varied in amount, and range from 2s. 6d. to 21s. per week for ordinary patients; domestic servants, when admitted on the recommendation of their masters, being charged a higher sum, varying from 5s. upwards. It is clear that at a cottage hospital, where the patients come from the immediate neighbourhood, no difficulty can exist in assessing the ability of the applicant to pay for hospital relief.

Where each person is known to the medical officer, or the vicar of the parish in which he or she may reside, a tolerably correct estimate of his or her means is easily arrived at. Under these circumstances, imposition, if attempted, is soon detected, and the would-be improvident are made to pay according to their means. We believe that the introduction of this system will eventually lead to important results. Experience, backed by the above figures, will sooner or later compel the managers of the large general hospitals to consider seriously, whether they are justified in continuing to administer so great an amount of free medical relief as they do at present. Amongst the artisans and the working classes generally, at the present day, admission to the hospitals for their wives, families, and themselves is looked upon as a right.

Free medical relief is not regarded by these classes as in any sense a degradation. On the contrary, although they would feel insulted if it were suggested that they ought by right to be regarded as belonging to the poorer classes, free medical relief has of late years been so rashly dispensed, irrespective of the ability of the patients to pay at any rate something towards their maintenance when in a hospital, that many of the lower middle class and well-to-do shop-keepers avail themselves of it, without a blush or any sense of shame. The classes relieved are not alone to blame for this demoralising state of affairs, for surely the competition between rival institutions to relieve the greatest number, and indeed an increasing number of patients every year, has had much to do with it.

Payments for medical attendance, on the provident dispensary principle, may be fairly regarded as likely to grow out of the system of patients' payments. For if the balance of income over expenditure continues steadily to increase, or, at any rate, to remain considerable, year by year, at a cottage hospital, we must express our strong opposition to the system of investing such balance as an "endowment fund." We venture to believe that the managers will see the desirability of still further fostering the feeling of self-help and independence amongst the patients, by giving a portion of their payments when in hospital to the medical staff. At Northampton nearly all the payments made to the Victoria Dispensary are divided

amongst the medical men, with the best results, the
expenses of managing the charity being defrayed by
voluntary subscriptions.

At Manchester, where theoretically the most perfect
system of provident medical relief ever tried in this
country has been attempted, about one-half of the
patients' payments are divided amongst the medical
staff. As a consequence of the provident system of
medical relief, the patients are much more independent
and self-reliant, pauperism is greatly diminished, and
great public good is accomplished. Every man is
taught under these systems, from childhood upwards,
that it is his duty to lay by something weekly against
the day of sickness, in order that he may choose his
own medical attendant, and pay him, like the best
of his neighbours, a fair fee for his attendance. We
should not be doing justice to the liberal spirit in
which the members of the medical profession have
from the earliest times given their services to the sick
and necessitous poor who find their way into our hospitals, if we did not state here that we are not advocating a system of small medical payments, at the
expense of the cottage hospital or its patients, for the
advantage of the doctor. Far from it. Before any
payment is made to the doctor, let every requisite be
obtained for making the hospital as complete and
efficient as possible; but, when once this desirable end
is accomplished, let not its managers set themselves
to accumulate invested property. Rather let them
remember that "the labourer is worthy of his hire,"

and that it is a plain duty they owe to the public to take care that the institutions under their management are not only made available for curing the sick, but that they are in addition fields of responsibility, which should be cultivated assiduously in the interests of the public weal. If the cottage hospital is made a place to which the sick but thrifty poor can resort in the hour of sickness, ought not every means to be taken to ensure during their residence there, that the feeling of self-respect which has led them to lay by something to pay for their treatment in hospital, shall be encouraged rather than effaced? The answer must necessarily be in the affirmative.

Let the managers of many medical charities ask themselves whether the indiscriminate free relief which they often give at the large hospitals of this country does not prove them to be "unfaithful stewards." We do not try to palliate; we desire to call attention to a plain truth. Let not cottage hospital managers follow so baneful an example, or the improvidence, dissipation, and disease, which are the disgrace of the poor in our large towns, will surely be extended eventually to country districts also. Let them rather endeavour, whenever the time arrives in the career of the institution under their management in which a continuous yearly surplus exists, that the provident principle may be further extended from patients' payments to the hospital, to patients' payments to the medical attendant. This can easily be done by setting aside a portion of those payments for distribution amongst the members of the medical staff. As a layman, we argue from sad experience in the past that this subject must not be allowed to be passed over in silence, but must be boldly faced whenever an opportunity is afforded in the history of a cottage hospital.

The medical profession has given its services gratuitously to the cottage hospital managers without hesitation, and in not a few instances members of the profession have been the originators and are still the chief supporters of many a cottage hospital. Let the managers show, by their adoption of the plan we so strongly urge upon them in preference to that of hoarding up surplus funds, that they are not unmindful of the principle on which we advocate this great change. As forerunners in the march of hospital reform, by the institution of the system of patients' payments as a special feature of their management, they have earned and will obtain the gratitude of all the more discriminating amongst the charitably disposed. Now, if they gradually introduce the payment of a fee, however small, to the medical attendant, in the interest of the public and not of the profession, they will still further increase this feeling of gratitude and admiration. The funds available will never be very considerable. nor in one sense adequate, for the average of patients' payments yearly at 174 hospitals was less than £60. Still the principle is important; and as the medical staff is usually limited to two or three members, if half the patients' payments were given in fees, it would exceed the usual rate of payments made to the parish doctor

4

under the poor law. At any rate the system of provident cottage hospitals here suggested is worthy of a trial. It will lead to the universal adoption of some such system, not only at cottage hospitals, but elsewhere. The cottage hospital managers have been the pioneers in the movement of payments by patients for their attendance. Let them push their ambition further, and aspire to the proud position of proving to the managers of general hospitals, that eleemosynary charity is dangerous to the poor, the public weal, and the public health. *Apropos* to the foregoing remarks is a case quoted by "a local practitioner" in the *British Medical Journal.* He states his case as follows :—

"For three years the town of B—— has had a cottage hospital ; for each patient admitted 3s. a week had to be paid, either by the patient himself, by some kind friend, or by the parish. The grievance is this : that poor people with broken and smashed limbs, who would under ordinary circumstances come under the care of the parish medical officer, are now taken to the cottage hospital, to which the parish pays 3s. a week for their maintenance ; and the result is (the medical attendance being obtained for nothing) the parish medical officer is 'done' (*sic*) out of the fees connected with such cases, which we know range from £1 to £5, and which are the proper perquisites of the medical officer,—and indeed these cases are the only ones for which he receives anything like a sufficient payment."

We are sure that cottage hospital managers will

bear us out in stating that this was happily an isolated case of hardship to the poor-law medical officer. The system complained of is now abolished ; but we retain the paragraph for the guidance of others elsewhere.

In the number of the same journal for 11th March, 1876, Mr. Richard Gravely, medical superintendent of the Lerwick Cottage Hospital, gives conclusive proof of this, and quotes the following extract from Mr. Napper's original pamphlet on the subject :—" Of the 77 paupers, 10 were cases of accidents and operation, for which the board of guardians paid the usual extra fees, amounting to £36. It is made a condition of admission, fully acquiesced in by the board of guardians, that the extra fee due for any such case of a pauper admitted, *shall be paid to the surgeon who would otherwise have attended it.*" This is the usual rule at all the best managed cottage hospitals, and it has worked well and been almost universally acquiesced in by the boards of guardians throughout the country. It was clearly an oversight that a similar rule had not been enforced at B——. This was conclusively proved by the fact, that the otherwise excellent rules and regulations of this particular hospital contained no allusion to pauper cases. We only refer to this case here to show how necessary it is at the outset to study the rules of other institutions, and to profit by their experience. It is clearly wrong, and quite foreign to the object for which cottage hospitals have been established, that pauper cases should be admitted, unless this excellent and necessary rule is strictly enforced.

It is customary, although the system of patients'
payments has nearly everywhere been adopted, that
each patient, except in cases of serious accident requiring instant admission, should produce a letter of
recommendation or a ticket from one of the subscribers. We would here say that we are opposed to
the system of tickets, on the ground that they act as
no practical check to abuse. They often necessitate
much suffering to a patient, who, having no friends, is
obliged to canvass amongst his wealthy neighbours
until he can obtain a ticket ; they add to the expenses
of management by increasing the outlay for printing ;
and they tend rather to destroy than to foster that
feeling of independence and self-help which, we take
it, the cottage hospital, with its patients' payments,
was specially fitted to promote. Surely the actual
bodily condition of the applicant ought to be the sole
passport for his admission to treatment.

In considering this, the so-called " Free system," it
should be remembered that half the funds are derived
from voluntary sources, which do not carry tickets or
any privileges with them. Of this class are " Hospital
Sunday," most donations, patients' payments, and miscellaneous receipts, making altogether about 52 per
cent. of the ordinary income available for cottage hospital purposes. There can be no fear of imposition, if
the patients' payments are enforced. The more this
question is examined, the more popular does it become.
The governors, who now have the privilege of recommending a patient for treatment, will lose nothing by its

adoption, for they will still have the assurance that any
patient sent to the hospital will, as under the existing
rules, be instantly admitted to treatment, should his case
require it. In all hospitals it must not be forgotten, that
the managers have to rely entirely upon the free gifts of
" Hospital Sunday " and other casual sources to meet
the cost of treating accidents and urgent cases, now admitted without ticket. The ordinary subscriber gets
his equivalent in tickets for any subscription he may
give, and if he avails himself of this *quid pro quo* to
the fullest extent possible, there is of course no balance
left for the extraordinary cases.

There is something almost humiliating to the charitable mind in this reflection : Although I nominally
give a free-will offering to the hospital when I subscribe
a guinea or so a year to its funds, yet when I take
tickets for my contribution, and thus purchase privileges
for myself or my dependants, I to that extent lose the
pleasure and moral personal benefit attending the contribution of him " that giveth and asketh not again."
Then, if a subscriber gives away a ticket to an un-
worthy object, without inquiry, he takes upon himself a grievous moral responsibility, which cannot
be too highly estimated. On the other hand, if he
subscribes to the hospital, and leaves the responsibility
of rejection in the hands of the managers, whose duty
it is to prevent abuse, he will relieve himself from the
trouble of personal inquiry into the circumstances of
the applicants, and will have at the same time the
satisfaction of knowing that the necessary relief is

administered to the really deserving without exception, and to the really deserving only. We can well
understand that before adopting any such radical
change in their system of management as is here advocated, the managers of hospitals will ask themselves
—How will this affect our income? We find from
experience that if a few subscribers have withdrawn
from a charity when this system has been adopted,
their loss has not been materially felt. The fact is, the
change attracts a large number of small subscriptions,
which were withheld under the ticket system, because
the amounts were not large enough to entitle the givers
to receive tickets for their money, and so they felt their
support to be incompatible with the old system.

When we consider the advantage the cottage
hospitals have over the large general hospitals, where
the free system has been adopted with success, the
ease with which inquiry can be made into the real
circumstances of the case and abuse at once detected,
and the large proportion of the annual income which
is derived from contributors who give freely and without any condition whatever, we arrive at this conclusion. An earnest consideration of this subject in all
its bearings must lead to the almost universal adoption
of the free system, in conjunction with a graduated
scale of payments for all who are able and who desire
to pay something for the relief they receive. So far
experience proves that such a system has not led in
a single instance to any but good results; and we fail,
therefore, to see why all hospital managers should

not adopt it. It is now the general rule that no case can be admitted unless, in the opinion of the medical officer, speedy benefit is likely to be derived from treatment in a hospital. This is undoubtedly a good rule, and although it throws additional responsibilities on the medical staff, chronic and incurable cases being thus rejected, we hope it may be more generally adopted.

It may be well to give here the form of recommendation most commonly adopted by cottage hospital managers. The following is the one issued by the committee of the Alton Cottage Hospital. Of course it would require slight modification, if the free system were to be universally enforced, although for all practical purposes it would stand whether the patient is introduced by a subscriber, or brought to the hospital by his friends. We have made a few suggested alterations to meet the altered circumstances, which are placed in a parenthesis in italics :—

LETTER OF RECOMMENDATION [*Introduction*], with which all Applicants must be provided, except in cases of severe accidents or sudden emergencies.

hereby recommend   , aged   . by occupation a   , residing   , as a proper person to be admitted into the hospital, and I consider h capable of contributing   per week towards h maintenance.

(*Signed*)

Dated   *Subscriber.*

The recommender is particularly requested to state the sum which he considers the patient or friends are capable of paying. The amount will vary, according to their circumstances, from 2s. 6d. to 8s. per week.

I hereby undertake to ensure the payment of the above-named sum of   per week during the time   is under medical treatment in the hospital; and I further undertake to remove h   when required to do so by the Directors, and, in the event of death, to defray all funeral charges.

(Signed)

Dated

The above must be signed by some responsible person.

Subscribers [*persons wishing a patient to be admitted*] are requested to communicate with one of the Directors, *before sending a patient to the hospital*, and, when practicable, to forward a statement of the case from the previous Medical Attendant.

No case can be admitted unless, in the opinion of the Medical Officer, speedy benefit is likely to be derived from treatment. Patients afflicted with mania, epilepsy, infectious or incurable diseases, are inadmissible.

It is requested that subscribers [*the persons sending the case*] will direct the patients to be sent as clean as possible, and with a sufficient change of linen.

The time has certainly arrived when it will be well for the general hospitals to begin at once to set their house in order with regard to the subject of patients' payments. There can surely be no objection raised by any one to the man who is unable to pay a heavy doctor's bill, with the necessary additional expenses of sickness, or to engage the services of a competent nurse, being admitted into the wards of a general hospital, the sole condition of his admission being that he contributes towards the expenses of his treatment according to his means. It is scarcely reasonable to condemn the philanthropist, who, moved by the pleading of an itinerant mendicant, puts his hand into

his pocket, and gives him a trifle, when the great
charities, with all the means at their disposal for easily
checking imposition, calmly allow their funds to be
annually spent upon the undeserving. Of course, the
direct result is that hundreds, nay thousands, of people
are annually pauperised by such a system of free medical relief, which any person can obtain by applying for it.
Such, however, is the practical result of the present
system of in- and out-patient relief in our large towns.

Many years' experience of its working compels
us to boldly state our belief that it is far too serious
an evil to be allowed to exist longer without a strong
protest on the part of the press and the public. The
cottage hospital managers, under great disadvantages,
have shown clearly enough how all abuse can be
avoided with a little judicious care. When the cottage
hospital system is more fully understood, we hope to
see some means adopted by which the general and
cottage hospitals may work together hand in hand,
with the twofold object of checking abuse, and affording
the best relief to the really deserving. It would surely
not be difficult for the county hospitals to make arrangements with the managers of the cottage hospitals
in their district, from whom all necessary information
as to the circumstances of the patients sent to them
from the country could easily be obtained, and a
system of patients' payments thus be established. We
unhesitatingly assert from actual experience, and we
appeal to cottage hospital authorities in general to
confirm the statement, that the really deserving are

only too glad to show their gratitude, by contributing
something to the funds of the charity to which they
owe so much.   Where the patients themselves are not
able to pay, their friends, more often than not, are
willing to pay for them.   We can go no further here
into this question of payments by patients in general
hospitals.[1]   It may, however, be added that the system
of payment by the poor-law guardians for paupers
when admitted to cottage hospitals is an excellent
one, and ought to ensure the universal adoption of a
similar rule by all the general hospitals in the country.
At any rate we are convinced that the evidence we
have adduced will suffice for the purpose of proving to
the satisfaction of all impartial observers, that the
thanks of the whole country are due to the originators
and founders of the cottage hospital movement, for
having initiated a basis upon which it will be possible
to erect a system of universal provident medical relief
to the exclusion of eleemosynary charity.

[1] In *Pay Hospitals and Paying Wards throughout the World*
the subject has been treated at length by the author.   A second
edition is now in preparation, and will be shortly published by
the Scientific Press, Limited, 428 Strand, London, W.C.

CHAPTER IV.

THE MEDICAL AND NURSING DEPARTMENTS.

Medical Department.—Advantages of Cottage Hospitals to country practitioners and their patients—Mutual advantages to consultants and general practitioners—Economy of time and labour of country doctors—Appointment of a medical director—Its advantages—Duties of medical director—Professional intercourse brought about by Cottage Hospitals—Cottage Hospitals as feeders to General Hospitals—Rules applicable to medical staff. *Nursing Department.*—Systems of nursing in force—Maintenance of regular supply of competent nurses—Cottage Hospitals must train their own nurses—Course of training for paying probationers—Advantages of training paying probationers—Annual reports should contain more details of the nursing department—Ladies' Committee—Rules for the Ladies' Committee of the Harrow Cottage Hospital—Advice and suggestions to nurses.

The Medical Department.

THERE cannot be, and, we believe, there is not at the present time in the minds of the medical profession generally, any doubt that the establishment of the cottage hospital has been in every way a decided advantage to the country practitioner, and to that portion of the public who reside in country districts. To the one, it has given increased experience and a

(59)

greater reputation amongst his patients. To the
other, it has secured the constant residence in the
vicinity of a class of professional men, who, thanks to
the cottage hospital, are able to bring in the hour of
sickness a skill and a ripe experience, which must tend
to strengthen the tie necessarily existing between
practitioner and patient. Nor do we consider that the
country practitioner and the country resident have
alone benefited by the cottage hospital movement ;
for we find that it is not an uncommon practice
amongst the staff of these small hospitals to invite the
surgeons of the county hospital to meet them in consultation, before deciding upon an operation in difficult cases. Thus the name of the town surgeon
becomes familiar throughout the county in which he
resides, his reputation is proportionally increased, and
his practice is thereby extended at a much more rapid
rate than formerly.

As a natural result of this state of things, country
people are beginning to have more faith in their own
surgeon, and are gradually being persuaded that, after
all, if one has to undergo an operation, it is much better
to have it performed at home, by one of the country surgeons, with all the advantages of pure country air and
complete quiet, than to secure the services of the most
eminent surgeon, if, in order to do so, it is necessary to
take up one's residence in London, with all its attendant
disadvantages. Although, at first sight, this may seem
to be likely to reduce the income, and to lessen the
reputation and fame of the urban surgeons, we doubt

if it will not rather prove the means of building up a
greater reputation amongst those very members of the
profession. The country surgeon will not feel justified,
even under the most favourable conditions, in taking the
entire responsibility of the difficult cases which occur
amongst the wealthiest of his private patients, without
first obtaining the opinion of some eminent specialist.
If mutual confidence and esteem are shown on all occasions by the consultant to the country practitioner, and
vice versâ, we do not hesitate to say that in the end the
profession and the public will alike benefit.

We hope, however, that the consultants in London
and in the large provincial towns will be very careful to guard and uphold the reputation of the country
surgeon. We regret to find that there is a growing
feeling amongst the profession in the county towns
and country districts, that it is scarcely safe to
send a really good patient to a London physician
or surgeon, because it so frequently happens, that,
instead of being examined, and the result of the
interview communicated to the medical man who sends
the patient in the first instance, patients often leave
their former attendant, and remain as the permanent
patients of the consultant. We have felt it our duty
to state this impression of the country practitioner
very clearly here, because we have been surprised and
pained to find this feeling so widespread and deeply
rooted. We believe, however, that it has its origin in
a misunderstanding on one side or the other ; and we
think that the mutual desire to benefit the patient,

which we are sure always exists on both sides, ought
to prevent the growth of such a feeling, and to create
one of mutual confidence and respect.

On another ground, the country practitioner is a
distinct gainer by the cottage hospital. It economises
his time and labour. The great difficulty with which
a country doctor has to contend is the wide area he
must cover in the course of one day's round. In bad
cases, when more than one visit is necessary in the
course of the day, the same ground has to be traversed
twice, however carefully the visits may be managed.
Now, however, it has become the rule at the majority
of cottage hospitals, to appoint one member of the
staff, either permanently, or (where more than one
practitioner resides in the village in which a cottage
hospital is established) two or more in rotation for a
certain number of weeks or months, whose duty it is
to look after all the cases in the institution that require
special care, and to give him the title of medical
director or medical officer. By this means, when a
severe case, requiring constant attention and frequent
visits, occurs in the practice of a country doctor,
the patient is as a rule removed to the hospital,
and thus all difficulty and anxiety of the specially
harassing kind before alluded to is obviated.

It will be seen from the foregoing remarks, that the
cottage hospital must be situated in close proximity
to the residence of the medical director having immediate charge of the patients, or many of the
advantages which it ought to afford to practitioner

and patient will be lost. In this way much time and labour will be economised, and from the constant supervision of the chief official, the utmost order, method, and regularity will be guaranteed. It might at first sight be thought that the selection of one out of several members of the medical staff for this specially honourable and responsible post would cause a feeling of jealousy or discontent amongst his colleagues. This is not found to be the case in practice, and a moment's thought will show the reason.

If some such arrangement for constant medical attendance were not provided, each medical man who had a case under his care in one of the wards, would be compelled to visit his patient, when seriously ill, two or three times a day, no matter how distant his residence might be from the hospital, and thus half the advantages offered to the profession by a well-managed institution would be lost. Where a medical director is appointed, he really is the house-surgeon for the time being, and discharges precisely the same offices for the patients under the care of the other members of the staff that the house-surgeon does for those of the honorary physicians and surgeons at the largest general hospitals. The duty of the medical director is not to initiate new systems of treatment for any but his own patients, but to carry out the wishes and views of his colleague, who would visit his own patients daily, and order any alterations which he might deem necessary or advisable. It will be seen that the duties of a medical director require considerable sacrifice of time, and the exercise of much tact on the part of the gentleman who holds the office. We think it would be advisable, where practicable, that the junior member of the staff should fill this appointment, because he would have more time on his hands, and, being fresh from the schools, in all probability he might be able to suggest valuable alterations in the regulations and arrangements of the cottage hospital. A precedent for this is afforded by the long-standing practice of the general hospitals in large towns, where it is the invariable custom for the junior member of the staff, whether physician or surgeon, to be appointed to the post of honorary secretary to the medical board, in which capacity he has to discharge all the routine work connected with the keeping of the minutes, the conduct of the correspondence, and so forth. In the cottage hospital he would, of course, be responsible for the proper keeping of the case-books, the ordering of drugs and other medical appliances, and, in the intervals between the visits of his colleagues, would act for them as their deputy, would carry out the details of treatment, and at the same time attend to the immediate requirements of all the patients. In this way, and under this system, much mutual confidence and good feeling have been engendered amongst the members of the profession, and an unanimous accord, on all important points, is thus secured.

Another advantage which the cottage hospital has secured to the profession in country districts is, that it has brought about a better feeling amongst all the medical men, as well as amongst the actual staff of the hospital, by affording neutral ground upon which all can meet for consultation and mutual intercourse, without loss of dignity, or any other disadvantage. In addition, an opportunity is offered to every practitioner, whether he be an honorary medical officer of the institution or not, to follow up his treatment with the aid of good nursing and a liberal diet,—for all members of the profession are at liberty, in most cottage hospitals, to treat the patients they may send to it, should they express a desire to do so.

On these points Mr. Edward Crossman, of the Hambrook Village Hospital, in his address to the governors, after some years' experience, said—" This professional intercourse is not the least valuable point of the cottage hospital system to the medical profession in the country. For the most part practising each in his own district, without much time for social intercourse, and accustomed to act upon his own judgment and responsibility, a feeling of distrust and jealousy too often springs up which in most cases only requires for its removal more frequent professional communication. The neutral territory to be found in the village hospital is the starting-point for that neighbourly feeling and action which promote the interest of the profession as much as that of the public ; and the consultation held over the hospital patient is often the commencement of cordial co-operation in private practice." This is valuable testimony from a practical man of considerable experience. When it is remembered how often jealousy and mutual distrust prevail amongst members of the profession in small country places, its value will be fully appreciated. Indeed, this testimony of their value to the profession at large ought to lead to the establishment of many other cottage hospitals throughout the country; for wherever an institution of this kind has been successfully started, there do we find the best possible feeling amongst all the members of the profession. The reason, of course, is that each member soon finds the value, nay, the necessity to himself, of his neighbour's co-operation in the more difficult cases, and this professional intercourse soon leads to more cordial relations in all respects. If the cottage hospital has only accomplished this, it will have earned its way to a place amongst the most valued of our public institutions.

Another point strikes us as being worthy of note here. It cannot be doubted that the cottage hospital not only saves many lives, and relieves the patient from much suffering, by placing a hospital within easy reach in severe cases of accident, but it further serves as a feeder for the general hospitals in large towns. We have met with many instances in which the patients have been sent by the staff of a cottage hospital to the larger county infirmary, because they desired a further opinion in a difficult case. Of course, here the town surgeon has the benefit of receiving a correct and detailed account of the patient's previous history, and the treatment which has been adopted, and he is thus enabled to form a more accurate opinion concerning the actual condition of the patient. This is no slight gain, and it should commend cottage hospitals to the support of the staff of the larger institutions.

The rules which are applicable to the medical staff are usually very brief, and the following may be accepted as embodying all that is included in the bye-laws of the majority of cottage hospitals :—

" No gentleman shall be eligible for the office of medical officer unless he be legally qualified, and duly registered, to practise under the Medical Act." "The medical officers shall be elected by the governors and shall have entire control over the medical management of the hospital." " They shall report to the committee, and shall act in concert with them for the good of the institution." " Any legally qualified medical practitioner residing in the district shall be allowed to attend cases sent by him to the hospital." " Each of the medical officers shall act as house-surgeon for a week (in rotation), and during such time shall take the general medical control, and in the absence of the other surgeons, shall, at their request, or, in cases of emergency, without it, attend to their patients." " The house-surgeon shall have power to admit any case at his discretion, and shall continue his attendance on all cases admitted during his week of office, unless such are sent to the hospital by another surgeon ; but each medical officer shall attend cases sent to the hospital on his own recommendation." "The surgeons are not expected to find drugs nor any surgical appliances, and they shall make application to the house committee for any articles they may require, which are not in stock."

The rules in force at Crewkerne are really excellent, and are well worthy the attention of cottage hospital managers. Much may be learnt by reference to them, and many valuable hints obtained from their perusal. On the other hand, the rules at some hospitals require revision, as, taking one example, we can hardly believe that any medical officer who gives his services gratuitously, will consent to be "removable by the committee *at their pleasure.*" On the whole, the rules in force at cottage hospitals, so far as we have been able to examine them, deserve commendation for the moderation and good sense evinced, and may be taken as evidence of the excellent understanding which exists between the lay committees and the medical officers.

The Nursing Department.

As regards nursing we have endeavoured to make our remarks as practical as possible, and with this object have divided the different nursing systems in force at the various cottage hospitals into two principal divisions. Thus we have as a first division those hospitals where there is one sole head over the nursing department answerable to the committee. This system is worked in three different ways in the cottage hospitals now open, *viz.* :—

(*a*) Where there is a lady superintendent or matron, with one or more nurses to assist under her.

(*b*) A lady superintendent or matron, with a nursing institution attached to the hospital.

(*c*) A head nurse in charge who has the entire control of the nursing and household arrangements, comprising the ordering of goods, provisions, etc., and the general control of the ordinary expenditure for hospital purposes.

In the second division we have the nursing arrangements and domestic management under the control of a ladies' committee, who visit the hospital regularly, and have complete control over the nursing staff, as well as over the general arrangements. Of this system we have three varieties, the two first of which are favourably spoken of, but the third is now considered obsolete.

(*a*) A working matron in charge of the nursing and household generally.

(*b*) A trained nurse in sole charge.

(*c*) A nurse who has to attend to the hospital when there are cases in it requiring attention, but when not employed in this capacity, she is allowed, and indeed expected, to visit and to nurse patients at their own homes. We may say of this last variety at once that experience has proved it to be impracticable ; and the late Mr. Napper, who was a warm advocate of it when he first started his hospital, became convinced that it is a system which cannot be made to work satisfactorily, and ought to be discouraged in the interests of the cottage hospital patients and of all concerned.

At several cottage hospitals special systems of nursing and domestic management have been tried, which may broadly be comprised and explained under the following two headings:—

(*a*) When the nursing and general management are undertaken by a sisterhood or church guild without any remuneration. This is the case at Middlesbrough (60 beds), Walsall (50), and Ditchingham (20).

(*b*) When the management in all respects is undertaken by a resident lady superintendent, who supervises all departments, and more often than not supports the hospital entirely or partially at her own expense.

It will thus be seen that almost every known system of nursing has been tried in cottage hospital management, and that several original schemes have been instituted also. We must next deal with the most difficult feature of this question,—how to maintain a regular and never-failing supply of thoroughly competent and efficiently trained nurses for cottage hospital purposes. In considering this question we cannot do better than glance for a moment at the systems in force at general hospitals, and see if the experience thus gained from the larger hospitals may not be turned to advantage in the case before us. We find by inquiry that the universal opinion in respect to trained and efficient nurses is this.—First, that at the present time the existing training institutions are relatively too few to enable them to supply the demands made of late years, by the general public, for trained nurses for private cases. If cottage hospitals, then, are to be supplied with a never-failing staff of efficient nurses they must train their own.

Of course, this statement was formerly met with the objection that the field being so limited, no scope is allowed to cottage hospital managers in this respect, and therefore such a system must be regarded as an impossibility. Experience, however, has largely negatived this contention. A trained nurse must for cottage hospital purposes be something more than a good nurse. She must be able to look after the housekeeping and domestic arrangements in the absence of the lady visitor, and must, in addition, be able to take charge of the hospital during the absence of this officer and the medical attendant. Here, then, separate and extra service is required, which will entail the exercise of much tact and patience, combined with good educational advantages. At every cottage hospital there is invariably a nurse, with more or less training, who is more frequently than not called the matron, and an assistant nurse, or assistant, or a probationer, whose duty it is to carry out the instructions of the nurse, and to help her, as she may direct, in the daily management of the hospital, and in the nursing arrangements. This system must be so arranged as to ensure that each cottage hospital matron may have under her direction, in the position of probationer nurse, a young woman of good abilities and education, who, with the assistance of the medical officer, can easily be made competent in two or three years to take entire charge of the cottage hospital, should anything happen to the matron or trained nurse.

Cottage hospitals, it is true, do not afford sufficient scope for training a large number of young women as nurses at the same time, but they afford an ample field for the establishment and successful working of a system like this. We are glad to notice that cottage hospital managers, for the most part, now train their own nurses. By such a system the homeliness of the cottage is secured, for the nurse is well known and respected by the patients. The former difficulty of finding the right person when suddenly required has thus been met, and a cottage hospital manager can recommend a nurse so trained to fill the vacancy when it occurs in the neighbouring hospital.

The system which we have been considering is one which we recommended fifteen years ago, before the training of nurses in cottage hospitals had in reality commenced. Since that time great changes have taken place, and it is now not uncommon for the committees—of the larger cottage hospitals, at any rate—to admit ladies to train as nurses for one year, and to give them a certificate at the end of that time which usually runs somewhat as follows :—

" This is to certify that   has received a year's training as a paying probationer in the   Cottage Hospital from   189   to   189   , and has been instructed in dispensing, the testing of urine,

and other practical duties appertaining to the office
of nurse."

These paying probationers are instructed in their
duties under the direction of the matron and of an
experienced staff nurse. They take it in turns to
act as surgery probationers, and are taught to clean
instruments, to prepare for operations, and to wait
upon the surgeons. They are also taught to dispense,
and under trained supervision they frequently make
up all the medicines used in the hospital. The
probationers attend lectures, including classes on
bandaging, conducted by the medical staff and the
matron, and have to pass a written examination
before they receive their certificates. The age at
which probationers are admitted to cottage hospitals
is usually from twenty-one to thirty-five years, a
certificate of health and testimonials of character in
a definite form being also required. They work
under the authority of the matron, and are liable
to dismissal by her at any time for misconduct or
inefficiency and neglect of their duties. The lady
probationers have to sign, at the end of a month's
trial, an agreement to complete their term of training,
and to pay a fee of from £6 to £10 per quarter in
advance, for which they receive board, residence and
medical attendance. Lady probationers have to provide their own uniform, and to pay for laundry.

At the Boston Cottage Hospital, Lincolnshire,
where this system has been in force for some years,
probationers' payments amount to between £70 and

£100 per annum, and the committee are enabled to
supply nurses to private cases in the neighbourhood
as occasion may arise. Such a system as this fulfils
a useful purpose, as it not only provides, as a rule,
a sufficient income to defray the matron's and nurses'
salaries, but it enables the lady probationers to readily
ascertain how far they are capable of becoming competent nurses, and prepares the way for their admission
to one of the larger nurse-training schools to complete
their three years of training and procure a certificate
which will constitute them trained in the fullest meaning of the term.

Despite the great advance which has taken place in
the quality and efficiency of the nurses engaged in
cottage hospitals, we regret to say that the annual
reports fail to give the necessary particulars of the
nursing department. Having regard to the circumstance that so many people take an active interest in
the work of cottage hospitals, it is not a little surprising that the reports should display, as a whole, so
small an amount of literary ability and prove so
meagre in information as to the work of the institution
to which they relate. If the honorary secretaries could
once be induced to attempt to make the annual reports
full and interesting statements of the work of each
year, in all its branches, they would find these reports
in great demand, and would have the satisfaction of
procuring a much larger volume of monetary support than they possess at the present time. A report
should give a full account of every branch of the

administration, and contain a frank statement of any
difficulties which may have been overcome. The
nursing of cottage hospitals is a subject surrounded
with interest. Its success means the removal of an
amount of unnecessary suffering amongst the poor,
too serious to think of with patience. By all means,
therefore, let there be an annual statement of the
progress and condition of the nursing department
of each cottage hospital. The omission of these
matters makes the reports of comparatively little value,
whereas their inclusion would make them of permanent value, and secure for them a popularity calculated
to materially aid the work in all its branches.

We have purposely omitted some pages of matter
formerly included in this chapter, which dealt at length
with the special circumstances of various so-called
"systems" to be met with in cottage hospitals. We
have taken this step because we are of opinion that
the great improvement in recent years in cottage
hospital nursing makes it desirable to attempt to
secure the employment of none but trained nurses of
the highest class in cottage hospitals, where skill and
training are of the first importance, owing to the smallness of the staff and the added responsibility which
devolves upon the nurse in consequence of the absence
of a resident medical officer.

We must now consider the constitution of the ladies'
committee. In every parish there will, we think, be
not the least difficulty in getting a sufficient number
of ladies to act. It must be an understood thing,

however, that they are to attend regularly, and to
supervise and advise the nurse in all domestic
arrangements, limiting themselves entirely to this
branch, or at the same time reading and giving instruction to such of the patients as may require it.
A list should first be made out of ladies willing to
act in this capacity, and they should then be requested to meet together and make such arrangements as may be convenient to them, by which one
of their number may be enabled to visit the hospital
daily in rotation, provision being made for occasional
absences from illness or other cause. Perhaps the
best arrangement in most cases will be that each
lady takes it in turn to attend daily for a week
at a time, the proper week being assigned to each,
whilst one of their number acts as a supernumerary,
to give extra visiting and assistance, or to take the
place of any other who may be prevented for a time
from acting. Some ladies may be disposed to interfere in the medical arrangements, but this action
must be immediately checked by the medical officer
before it assumes any shape, and such ladies must be
given to understand that they are to limit themselves
strictly to domestic arrangements. If any such tendency be quickly nipped in the bud, there will probably be no difficulty in the matter. In all cases the
lady of the week must be accountable to the committee of management for the goods, etc., ordered by
her during her term of office.

We give below the rules of the Harrow Cottage

Hospital with regard to the ladies' committee, which are very fully drawn out, and are almost identical with the plan here recommended.

(1) This will consist of seven ladies. including one to be specially elected by the board of managers to be lady superintendent.

(2) Their duties will be : —

(*a*) To superintend the domestic arrangements of the hospital.

(*b*) To give out the stores, and make recommendations as to the purchase of the same.

(*c*) To examine the household accounts. and present them to the treasurer from time to time for payment.

(*d*) To advise the board of managers on all points relating to the comfort of the patients.

(3) The lady manager will act as the representative of the ladies' committee in the ordinary routine of the above duties, and shall in addition :—

(*a*) Act as secretary of the ladies' committee.

(*b*) Fix in conjunction with the secretary of the board of managers (but subject to appeal to the board) the sum to be paid by each patient.

(*c*) Collect the weekly payment from patients, and pay the same periodically to the treasurer.

(4) The committee will meet not less than once a quarter, will keep minutes of all their proceedings, and report from time to time as required by the board of managers.

(5) In case of the temporary absence of the lady manager, the committee may appoint from amongst themselves a deputy to perform her duties.

We might suggest that to rule (2) should be added—

(*e*) To read to and instruct such patients as may be in a fit state of health.

It will be observed that we have omitted the classes

into which were formerly divided the nurses employed
at cottage hospitals. They were respectively—(1) the
trained nurse from some institution ; (2) the married
couple with or without children ; (3) the woman of
the village who has had some experience in nursing, or
who has been sent to a hospital for a time to acquire
it ; (4) a good nurse or assistant nurse, chosen from
the county hospital. The modern system of training
nurses, and the change in public and medical opinion
in regard to nursing, have had the effect of gradually
eliminating the differences which formerly existed ; and
we may now conclude that every well-managed cottage hospital is placed in charge of a thoroughly
trained and certificated nurse. In such circumstances
it would be superfluous to republish the remarks made
about each of the four classes above enumerated, seeing
that only the first is recognised as being fitted to
adequately discharge the duties required of her, if the
work of a cottage hospital is to be properly done.

We may perhaps usefully address a few practical
hints to all nurses who may be engaged in cottage
hospitals.

First of all, let us impress upon the nurse, whether
she be matron, trained nurse, or probationer, the
absolute necessity for great self-control, self-respect,
and self-reliance. Without these three qualities a
trained nurse in charge of a cottage hospital would be
almost as much out of her element as a bull in a china
shop ; for with the great responsibilities which must
perforce devolve upon the nurse in these small insti-

tutions, unless she be ready to meet all emergencies, to successfully grapple with difficulties, and to have perfect confidence in her own powers, she ought never to have taken office in a cottage hospital.

Our advice to her would be as follows :—Acquire regular and punctual habits, avoid bustle or noise in the quick and methodical discharge of your daily duties, and above all things be sure to be neat and orderly in matters relating to your own dress and appearance. Always strive to anticipate the wants of the surgeon, learn his peculiarities in the way of treatment, and endeavour by a careful study of each case to acquire a habit of proper observation. By this means only will you be able to properly place before the surgeon on his rounds an accurate report of the progress of the case. Be firm but tender and considerate to your patients, and courteous to your superiors in all things, and on all occasions. When a visitor calls at the hospital, be he poor or rich, known to you or a stranger, always receive him with frankness, and be careful to answer any questions concerning the hospital with prompt and cheerful civility.

CHAPTER V.

DOMESTIC SUPERVISION AND GENERAL MANAGEMENT.

Ventilation—Necessity of constant interchange of air—Daily inspection by head nurse—Flowers—Regularity in administration of medicines and of food—Diet—Family prayers—Bathing patients—Bed-making—Draw sheets—Patients' clothes—Uneaten food—Bed sores and their treatment—Administration of cod-liver or castor oil—Preservation of ice—Administration of alcohol—Treatment of accidents on admission—Treatment for a sprained ankle—Enemas—Poultices—Hot fomentations—Administration of food or medicine to semi-conscious patients—Fainting—Fits of different kinds—Frost-bite—Foreign bodies in ear, etc.—Dressing burns—Hæmorrhage—Injuries to head—Operation room—Sponges.

IN the pages which compose this chapter no attempt has been made to exhaust the subject of domestic supervision or management of a hospital, since that would require almost a volume to itself. It is hoped, however, that the following notes, disjointed though they may be, will be of some service and profit to the reader.

Ventilation.—In the majority of the cottage hospitals, especially where a building has been adapted for the purpose, the chief means of regulating the

(80)

ventilation of the rooms will be by the windows. It is by means of the windows that the greatest amount of ventilation will be obtained, if judicious care be exercised. The nurse should always bear this in mind, and remember the great importance attaching to the free circulation of pure air throughout the wards. She should, in spite of the remonstrances of her patients, who are sure to object strongly to the admission of fresh air, take care while keeping the temperature of the ward at from 60° to 65°, and never allowing it to exceed 70° (unless specially instructed on this point by the medical man in charge of the cases), that a constant interchange of air is, without fail, at all times kept up in every room.

We were much struck with the pure atmosphere and airy freshness of the wards at Cranleigh Village Hospital, where, under great natural disadvantages, by impressing upon the nurse the necessity of keeping the windows *always* open about an inch or two inches at the top, almost perfect ventilation has been secured in small wards, in which the cubic space is far from ample, and where the structure of the building would, at first sight, seem to prohibit so great a desideratum as the constant supply of fresh air. Nurses must remember that in their vocation, above all others, it is of no use working by fits and starts, but with regularity and method. Let them apply this to the ventilation of their wards, and by avoiding the popular error of opening the windows top and bottom for an hour or so each day and then closing them till the next

morning, and by insisting upon a constant supply
of air being admitted to the wards through the
space left at the tops of all the windows, profit
by the hints we have here given on this important
point.

Daily Inspection.—It is very desirable that the head
nurse should make a complete inspection of the whole
of the hospital buildings at least once a day, and that
she should more frequently inspect the lavatories and
water-closets, to see that strict cleanliness is observed
everywhere, and that the proper amounts of disinfectants are being used.   Care should be taken not to allow
slops or other matters to be thrown down the closets,
and if, as we advise, earth-closets are more universally
adopted, this practice will, of course, be practically
impossible.   Still it is necessary to caution the nurse
against the practice, which is far too prevalent, of
allowing things of any description to be thrown down
the closets.   We have seen scrubbing brushes, tow,
lint, poultices, flannel, wearing apparel, linen, knives
and forks, and indeed almost everything, possible and
impossible, discovered as the cause of an overflow into
the ward owing to stoppage in the pipes connected
with the water-closet.   Have a proper place for
everything, and a proper receptacle for all waste
material, for by this means the wards will present an
appearance of neatness and comfort which nothing
else will ever give them.

Flowers.—In the summer, when possible, be sure to
secure a constant supply of fresh flowers, which will

add much to the cheerfulness of the wards, and to the comfort of everybody. In the winter a carefully-selected bouquet of everlasting flowers presents a pleasing and effective appearance.

Regularity in the administration of the medicines, and of the food ordered for the patients, should always be observed. A good nurse is known by the exact and faithful way in which she carries out the instructions given her by the medical officer. On the strict observance of these apparently trifling, though really very important details, the life of the patient and his ultimate recovery not infrequently depend. The nurse should avoid familiarity with her patients, and be scrupulously modest in all her dealings and intercourse with the male patients. *On her* the happiness and in no small degree the success and popularity of the hospital will depend ; she should take care, therefore, that she is in all respects worthy of her honourable calling.

Diet.—It is a fact worthy of record, that in many instances no diet table or scale of diets is given in the printed reports of cottage hospitals. Cottage hospital authorities should have a fixed scale of diet ; for proper diet—good wholesome food in right quantities and of the most suitable kind—is one of the great features of successful hospital treatment. In the hope that they may prove useful the following three diets are offered, not necessarily for adoption but as a guide to the profession in this matter :—

Fever or Low Diet.

Bread	. . .	6 oz.
Milk .	. . .	2 pints
Beef Tea or Mutton		
Broth (1 lb. to oj.)	.	1 pint
Arrowroot .	. .	2 oz.
Tea .	¼ oz. } for a pint	
Sugar	1 oz. } of tea.	

Middle or Half Diet.

Bread	. . .	8 oz.
Butter	. . .	1 oz.
Fish	. . .	6 oz.
or Hashed Mutton	.	3 oz.
Potatoes, mashed	.	8 oz.
Milk	. . .	1½ pint
Rice, Sago, or Arrowroot	2 oz.
Tea	. ¼ oz. } for 1 pint	
Sugar	. 1 oz. } of tea.	

Full or Ordinary Diet.

Bread .	.	16 to 20 oz.
Meat (cooked without bone)	. .	7 oz.
Potatoes	. .	12 oz.
Butter .	. .	1 oz.

Cheese, or Gruel	.	2 oz.
Milk	. . .	½ pint
Tea	. . .	1 oz.
Sugar	. . .	1 oz.

Breakfast (7 a.m.)	*Dinner* (noon)	*Tea* (4 p.m.)	*Supper* (8 p.m.)
Bread 4 to 6 oz.	Meat cooked 7 oz.	Bread 4 to 6 oz.	Cheese 1 oz.
Butter . ½ oz.	Potatoes 12 oz.	Tea . ½ pint	Bread 3 oz.
Tea ½ pint (with Milk and Sugar).	Cheese 1 oz.	Butter ½ oz.	or Gruel or Oatmeal.
	Bread 4 oz.		

½ oz. of cocoa might be given for breakfast instead of tea. Rice pudding might be given for dinner alternately with bread and cheese. Eggs, poultry, beer, wine, and spirits, to be ordered by the medical officers. Condiments at discretion. Should Liebig's Extract be used for making beef tea, that made from the meat itself must be also made use of in turns with it. Gifts of vegetables, poultry, game, etc., will help to vary the diet.

Family Prayers.—With the consent of the chaplain and committee, we would suggest that each head nurse should every morning read prayers to all the convalescent patients and the assistants under her in the convalescents' sitting-room. This is an excellent plan, as it infuses a good moral tone into the whole establishment, and there is a great comfort

and benefit to be gained by commencing the day
well. Besides, the cottage hospital presents great
opportunities for doing good to many a wandering
soul, and what can be more beneficial to nurse and
patients, surrounded as they frequently are by the
dying or the seriously wounded, than the daily assem-
bling together before the labour and trials of the day
begin, to offer up praises and prayer to Almighty God
for all His benefits and loving-kindness? Depend
upon it that such a line of conduct as is here suggested will only lead to good results, and will, if
carried through with tact, be found a great comfort
and benefit to all concerned.

Bathing Patients.—It is very necessary that patients
should be kept perfectly clean, from the day they
enter to the day they leave the hospital; and it
should be the invariable rule to insist upon every
patient taking a bath before he is put to bed on his
first arrival at the hospital, unless there is some
special circumstance which prevents such a course
being pursued. In the case of a patient whom the
doctor has ordered the nurse not to bathe, she should
take the earliest opportunity, before he is allowed to
be put to bed, to wash him as thoroughly as possible
with warm water and soap. Unless this rule is
strictly enforced it will be found impossible to keep
the wards clean and free from many objectionable
features, while the beds will soon become more lively
than habitable. This, of course, will be found a difficult duty to efficiently discharge; but with firmness

and tact it will eventually be got over. So deeply rooted is the prejudice to soap and water inherent in certain of the poor classes of our population, that we have known patients more than once (in each case a woman) leave the hospital and reject all treatment rather than submit to be placed in a hot bath. This strong prejudice must be overcome at all hazards, and when once a patient is admitted to the ward, care must be taken that he frequently (two or three times in each day) washes his hands, and that he has a bath at least once a week. With such rules as these the patients will be happier, and the wards will be a credit instead of a disgrace.

Bed-making plays an important part in the daily routine of a nurse, as upon the careful arrangement of the pillows, sheets, and bed, depend in no small degree the comfort of a patient who is really ill, and his freedom from bed sores and like evils. When a nurse has a bad surgical case, with intermittent hæmorrhage or profuse discharge, she must be careful to keep a clean mackintosh constantly under the wound, and to place a draw sheet over it. By this means it will be easy to keep the patient clean and sweet, nor will it be difficult to frequently change the draw sheet as occasion may require. When it is necessary to change the bed-clothes of a bed-ridden or nearly helpless patient, the following will be found an easy course to pursue. Having the clean sheet ready, roll up the dirty under sheet as close to the patient as possible, then half roll up the clean sheet, and place the unrolled half over that portion of the bed from which the dirty linen has been removed. Then lift your patient on to this, and having removed the remainder of the dirty sheet and replaced it by unrolling the clean one, the patient will be made comfortable very rapidly, and with the least possible inconvenience. If the patient be too weak to be moved bodily as we have suggested, it is not difficult to change the under sheet without lifting the patient much, providing the aid of an assistant is secured. With this method it is necessary to begin at the head of the bed, to gradually withdraw the dirty sheet and at the same time to replace it with the clean one, which must be rolled up and put in readiness at the head of the bed before the dirty linen is removed. With a little practice it is not difficult to do this quickly and without any discomfort to the patient.

In surgical cases, fractures, and the like, the patient will be able to grasp the bed-pull, and thus raise himself sufficiently to allow the sheets to be changed without any trouble or delay. It is unquestionably an advantage to be able to change the bed linen without bodily lifting the patient as the first plan necessitates, and we therefore recommend, for general adoption, the latter one in preference. The *draw sheet* is one of the most serviceable agents in the nurse's hands, to secure the double purpose of keeping her patient dry and protecting the bed. Her great object should be to keep her patient on a clean, dry sheet. Sometimes, as after lithotomy and other operations, the discharge

is so constant that the sheet requires changing very
frequently, and it is of the utmost importance that this
be done with the least possible disturbance.   A soft
old sheet having been folded to the required width
(two feet will generally be found sufficient), let the sheet
be rolled up at one end, leaving just sufficient of it to
pass under the patient's buttocks.   When the sheet is
wet draw it through from the side opposite to the one
from which it was first passed under.   To do this
unroll just enough of the clean end to secure a dry
piece under the buttocks.   The soiled end may then
be rolled up tight and pinned.   In this way one draw
sheet will be sufficient for several changes, and by
pinning to it a clean one, a succession of draw sheets
may be passed under a patient with a minimum of
disturbance.

Patients' Clothes, and Uneaten Food.—No patient
should be allowed to have his clothes anywhere near
or about his bed. A proper locker should be provided
for them, in which, when neatly folded, they ought to
be kept. Unless the nurse is careful to insist upon
the observance of this rule, her wards will always
have an untidy appearance. At many hospitals it is
an invariable rule to give the patient a suit of ward
clothes, made out of old blankets or flannel, in which
he is sent to the ward after having his bath on admission. All his clothes are then removed to the disinfecting closet, and are thoroughly fumigated and cleansed
before they are allowed to be taken to his locker in
the ward. When possible, this course should always

be taken, and in exceptionally bad cases, the clothes should be destroyed and a new suit supplied to the patient. In the evening, before turning out the lights, the nurse should carefully remove from the patients' cupboards all unconsumed food which she may find there, as it is not an uncommon case for patients to hide up their food, or a portion of it, for their friends. The writer has known instances of patients suddenly manifesting symptoms of *delirium tremens*, although they have been in the hospital for some weeks, when on inquiry it has been found that the man had saved up each day's allowance of brandy until he had secured a good supply, when he availed himself of the earliest opportunity of taking the whole quantity at one time. But apart from this, it must be remembered that food kept in the wards soon becomes contaminated with impurities to a greater or less extent, and hence the necessity of giving the patients their meals or diet at fixed times, and the advisability on this ground also of allowing no food to remain in the ward throughout the night. It is impossible to hope for the recovery of a patient unless the orders of the medical attendant, in respect to both diet and medicine, are exactly obeyed.

Bed Sores.—Every experienced nurse is familiar with that most troublesome of pests, the bed sore. It will therefore be well to give a hint or two as to the steps which it is advisable to take to avoid them breaking out, where the patient has to submit to a long treatment in bed. It should always be borne

in mind that they are caused by allowing the patient
to lie for a long time in one position, and thus producing continued pressure in one or two parts of the
body. Very slight changes of position will do much to
relieve the patient. It is not always possible to prevent the sores, but as a rule they do not occur to
patients who have the good fortune to be nursed by an
experienced and careful woman. The first point to
observe is perfect cleanliness, smoothness, and dryness,
the sheets being always changed *at once* if the least
damp or wetness is discovered. By the skilful arrangement of pillows, and the timely use of air cushions or
water pillows, much can be done. It is not a bad plan
to apply to the exposed parts, which should be lightly
and carefully rubbed, an application of collodion, or a
wash composed of two grains of perchloride of mercury
to an ounce of spirits of wine, or some spirit (whisky
or brandy for preference) and water.[1] *Emplastrum
elemi* (Southall's) spread on white leather is an excellent application. If these precautions are taken,
and the earliest symptoms of the appearance of a
sore treated with care, the nurse may more often
than not save the life of her patient, and thereby

---

[1] The following is a very good lotion for painting over a *red-
dened* patch of skin before it has broken, with the object of preventing a bed sore. The tannin in the catechu acts like the
perchloride of mercury, but the lotion possesses the advantage
of not being so poisonous. Forming also a thin paste, it coats
the skin with a slight film, thus still further protecting it.

℞ Liq. Plumbi Subacet. Dil.

Tinct. Catechu āā partes æquales. M. ft. lotio.

do greater justice to the treatment of the medical
officer.

To give Cod-liver or Castor Oil.—Much trouble is
caused to the nurse at times by the obstinate refusal
of patients to take the medicines ordered. Especially
do they object to swallow oil of any kind. The
best way we know of taking oil is to rinse out a wineglass with a little brandy, whisky, or other spirit,
leaving one drop of the spirit at the bottom of the
glass. Having done this, pour the dose of oil into
the wineglass, and the spirit will roll the oil, so to
speak, into a ball, like the yolk of an egg, which can
be easily swallowed *en masse*, without any unpleasant
taste. A little milk taken immediately afterwards
will be found useful. Another good plan is to divide
a lemon, squeezing the juice from each half into
separate tumblers. To the one add about a wineglassful of water and sufficient sugar to make it palatable.
In the other tumbler beat up the dose of oil with the
lemon juice, then add some sugar and a little less
than a wineglassful of water ; stir this well up to the
moment of swallowing, then give the patient the
previously-prepared lemonade. When this has been
taken, it is pleasant to wipe the teeth with the inside
of one of the lemon halves previously used. Other
writers have recommended that the oil should be
mixed with milk or coffee, or it can be carefully
mingled with a small basin of soup, and this latter plan
is often found useful. For our own part, having been
compelled to take large quantities of cod-liver oil

from time to time, we can confidently recommend the adoption of the first plan, because it will ensure perfect freedom from any unpleasantness of taste or after effects.

Preservation of Ice.—The preservation of ice is often a very difficult though important matter in dangerous cases, and we make no apologies for quoting the following excellent paper by the late Mr. Sampson Gamgee, F.R.S.E.[1]

" The luxurious comfort and practical benefit which many patients derive from the frequent ingestion of ice are well known. To those who can command a constant attendant, and to whom cost is no considera-tion, it is comparatively easy to secure a constant supply of little lumps of ice at the bedside ; but even under these favourable circumstances, it not unfre-quently happens that in the warm bedroom, towards the small hours of the morning, when the bit of ice is most wanted to suck, or to be put into milk, water, or other beverage, it is found to have melted, and time is lost, and perhaps the household disturbed, before a fresh supply can be obtained. This is frequently the case when the lumps of ice broken for use are kept in a glass or saucer in the room. My practice for some years has been to cut a piece of flannel, about nine inches square, and secure it by ligature round the mouth of an ordinary tumbler, so as to leave a cup-shaped depression of flannel within the tumbler to about half its depth. In the flannel cup so constructed,

[1] See *Lancet*, vol. i., 1876, p. 846.

pieces of ice may be preserved many hours, all the
longer if a piece of flannel from four to five inches
square be used as a loose cover to the ice-cup. Cheap
flannel, with comparatively open meshes, is preferable,
as the water easily drains through it and the ice is
thus kept quite dry. When good flannel with close
texture is employed, a small hole must be made in
the bottom of the flannel cup, otherwise it holds the
water, and facilitates the melting of the ice, which is,
nevertheless, preserved much longer than in the naked
cup or tumbler. In a room 60° Fahr. I made the
following experiment with four tumblers, placing in
each two ounces of ice broken into pieces of the
average size for sucking. In tumbler No. 1 the ice
was loose. It had all melted in two hours and fifty-
five minutes. In tumbler No. 2 the ice was suspended
in the tumbler in a cup, made as above described, of
good Welsh flannel. In five hours and a quarter, the
flannel cup was more than half filled with water, with
some pieces of ice floating in it ; in another hour and
a quarter (six hours and a half from the commencement of the experiment) the flannel cup was nearly
filled with water, and no ice remained. In tumbler
No. 3 the ice was suspended in a flannel cup made in
the same manner and of the same material as in No.
2 ; but in No. 3 a hole capable of admitting a quill
pen had been made in the bottom of the flannel cup,
with the effect of protracting the total liquefaction of
the two ounces of ice to a period of eight hours and
three quarters. In tumbler No. 4 two ounces of ice

were placed in a flannel cup, made, as above described, of cheap open flannel (10d. per yard), which allowed the water to drain through very readily. Ten hours and ten minutes had elapsed before all this ice had melted.

"A reserve supply outside the bedroom door can be secured by making a flannel cup, on the plan above described, in a jug, and filling it with little lumps of ice, care being taken that there is space enough below the bag to allow the water to collect, and leave the ice dry. This provision will allow the ice to be used during the hottest night, without the supply failing, or the patient being disturbed—two very important considerations. The real therapeutic benefit of ice is only produced in some cases by its free use, and its soothing and stilling effect must be aided by the most perfect surrounding quiet."

A long and very agreeable association with the late Mr. Gamgee at the Queen's Hospital, Birmingham, taught us to value highly his ripe experience and ready ability in cases of emergency. Mr. Gamgee fully recognised the importance of trifles, and through his courtesy we have been enabled to publish many similar hints on apparently trivial though really important matters of detail.

Administration of Alcohol.—We do not feel called upon to make any extended comments on this subject; but we append a table showing the amount annually expended at 18 London general hospitals and 21 cottage hospitals, from which it will be seen that the

cost of stimulants per head per patient in the metropolitan hospitals is 2s. 8d. and in cottage hospitals
2s. 9d.

THE ADMINISTRATION OF ALCOHOL.

*Table showing the consumption of Wines, Spirits, and Malt
Liquors, at 18 Metropolitan General and 21 Cottage Hospitals,
with the number of In-Patients to whom they were administered, for the year 1893.*

Year 1893. Metropolitan Hospitals.	In-Patients.	Cost of Stimulants.	Year 1893. Cottage Hospitals.	In-Patients.	Cost of Stimulants.
London - - - -	10,599	£1,600	Ashburton - - -	49	£6
Guy's - - - -	6,160	215	Batley - - - -	162	17
St. George's - - -	4,700	1,051	Bourton-on-the-Water	45	7
Middlesex - - -	2,513	440	Bromley - - -	175	24
St. Mary's - - -	4,274	519	Bromsgrove- - -	106	7
King's College - -	2,566	457	Budleigh Salterton -	53	3
University - - -	3,228	260	Chalfont St. Peter -	11	4
Westminster - - -	2,950	287	Egham - - - -	117	16
Charing Cross - -	2,503	330	Grantham - - -	199	68
Royal Free - - -	1,554	157	Hambrook - - -	50	11
Seamen's - - -	2,582	428	Hammerwich - -	77	14
German - - - -	1,456	402	Iver - - - -	60	2
Great Northern Central	1,220	75	Lytham - - -	91	3
West London - -	1,611	181	Newton - - -	189	3
Metropolitan - - -	774	78	Petersfield - -	63	13
Poplar - - - -	443	52	Retford - - -	22	3
London Homeopathic -	445	70	Romford - - -	67	2
Miller - - - -	203	51	Shepton Mallet - -	150	6
			Tamworth - - -	200	51
			Tetbury - - -	55	10
			Wellington - - -	45	5
Total - - -	49,781	£6,662	Total - - -	1,986	£275

Summary.

                                                                s. d.

Average cost per head per patient at 18 Metropolitan Hospitals, 2 8
    Do.   do.   21 Cottage Hospitals, . 2 9

The Treatment of Accidents on Admission.—It will
often happen, especially in the experience of cottage
hospital life, that no one but the nurse will be immediately available when a case of severe accident first
arrives at the hospital. It may, therefore, be well to
give a few plain directions as to the course to be pur-

sued in such an emergency. Every cottage hospital
should be provided with a stretcher made of canvas,
with strong loops on either side, through which the
poles may be placed and withdrawn at pleasure. When
a patient is brought to the hospital, he should be carefully carried in on the stretcher, and placed on the
couch in the accident room, when the poles can be
removed. Great care should be exercised in the case
of a broken or fractured limb, to prevent the serious
additional accident which, by careless carrying, has
often happened, of converting a simple into a compound
fracture by forcing the fragments of bone through the
skin. To prevent this, the nurse should take charge
of the broken limb, and superintend its movement.

Before the doctor arrives, in cases of great ex-
haustion a little brandy and water should be given
to the patient: but it is only too often the case,
that the sufferer is brought in, reeking with brandy,
given him most recklessly by his comrades. Care
should be taken to loosen all the clothes about the
neck, the necktie and collar being carefully removed.
The boots should be pulled off, or, in the case of an
injury to the foot, the boot may be cautiously cut away,
and thus all unnecessary suffering avoided. The next
step is to wash the patient as thoroughly as practicable,
after removing most of his clothing. In the case of an
injury to the leg or thigh, the trousers and stockings
should be slit up the seams, and thus freed from the
wounded limb. Now prepare whatever is requisite in
the way of appliances for the treatment of the patient—

in the case of fractures : splints, pads, bandages, cotton wool, and the like ; in the case of wounds : lint, bandages, picked oakum, sticking-plaster, etc.—and have ready for use the necessary instruments, with plenty of hot water and sponges. In this way much time will be saved and the patient will, with the least possible delay, be ready for removal to the ward, the poles being replaced for this purpose, and the bed being previously prepared for his reception, with the necessary waterproof sheetings, draw sheets, pillows, and other requisites.

Treatment for a Sprained Ankle.—On this point the following observations of the late Sir Erasmus Wilson may be of value :—" We all know that there is nothing more painful than a sprain of an ankle ; it will lay a man up longer than the fracture of a bone, and he may recover with a very weakened joint. Accompanying a country medical man on his rounds, he told me he had made a great discovery in the treatment of sprains. ' The way I cure a sprain,' he said, ' is this : I take some warm lard ; I warm it, and rub it into the sprain half or three-quarters of an hour. I then take some cotton wool and wrap around the joint, and put on a light bandage. The sprain, which would have taken many months to get well, gets well in a few days— certainly in a few weeks—without any ill effects or after-consequences.' " Sir Erasmus Wilson adds : " I tried this treatment, and found that it succeeded admirably."

Practical Points.—This chapter may appropriately be concluded by giving a few plain practical hints for

7

nurses, regarding the proper carrying out of some of their duties, and the immediate treatment to be adopted in a few cases of emergency, likely at any time to arise. There are, firstly, three duties, concerning which little can be said, but every nurse should endeavour to make herself thoroughly acquainted with them. These are: —(1) the application of leeches and blisters; (2) the passage of the female catheter; (3) the mode of using the clinical thermometer.

To give an Enema.—In the administration of enemata, the nurse must take care first to well oil the tube, before introducing it, and also to pass through it a few syringefuls of the liquid, not only for the purpose of ascertaining that the instrument is in good working order, but also that the whole may be well warmed before introduction, and that any air may be expelled. Enemata are given with two objects :—(1) to remove fæces from the lower part of the intestinal canal, when soap and water, gruel, etc., are used, and the injection must be a large one (1½ to 2 pints or more); (2) as a means of supplying nourishment to the patient, when, as the object is to retain the nourishment, very small injections must be given (not more than 2 oz.), and a small elastic bottle of that size may be used.

Poultices.—The use and comfort of poultices depend greatly on the manner in which they are made and applied. For a *bread poultice* use stale bread. Place the required quantity in a basin, pour over it sufficient boiling water to soak it, and let it remain for about five minutes with a plate covering the basin, then drain off superfluous water, and place the bread between layers of muslin or soft old linen for use. *Linseed poultices* are too often made with powdered linseed from which the oil has been so thoroughly extracted as to make it as dry as sawdust. Crushed linseed is the right material, and one good test of its nature is its greasing the paper in which it is wrapped. First scald the basin in which the poultice is to be made, then put into it a quantity of meal suited to the particular case. Any knots or clusters of meal are to be crushed with the hand. Some well-powdered charcoal may be added to the meal when the poultice is to be applied to a part discharging offensive matter. A wooden spoon or spatula is the best instrument for mixing and spreading with the least loss of heat. Boiling water should be slowly added whilst the meal is constantly stirred, till it is of proper consistence. One great fault is that of adding too much water. When sufficiently mixed turn out the poultice on a piece of lint or clean rag, over which it is to be evenly spread, the sides of the rag being neatly folded into the margin of the poultice, and a piece of thin muslin placed over it to come next the skin and admit of easy and clean removal in due time. Tow finely teased out answers as a cheap material on which to spread poultices in hospital. When a large linseed meal poultice has to be applied to the abdomen or to the front and back of the chest it may be made in a wash-hand basin and spread on a chamber-towel. A soft old handkerchief in the absence of muslin may be laid on the poultice. A poultice thus made may be rolled up and carried from a downstairs kitchen to a top bedroom with so little loss of heat as to require caution in its application for fear of scalding.

For *mustard poultices,* the essential is mustard of first-rate quality. Mustard leaves are a very good substitute.

Hot fomentations to be comfortable and useful must be hot, and the cloths well wrung out, otherwise the superfluous water soaks the patient's clothes and the bed, and causes much annoyance. New flannel is the best material for fomentation cloths. Put a towel over an empty basin, the dry flannel in the towel, and pour over it boiling water, for good soaking. Quickly wrap the towel round the flannel, and twist the two ends of the towel in opposite directions, so as to squeeze out all the water. A perfect wringing machine may be quickly made by loosely stitching in the two ends of the towel round pieces of wood,—a walking-stick cut in half answers the purpose perfectly. When the steaming flannel has been quickly applied, cover it and the adjoining parts with a piece of waterproof sheeting, which has been previously warmed enough for comfort. This is also a very useful covering for large linseed meal poultices.

The administration of food or medicine to a patient in a semi-conscious condition is often a matter of some difficulty. The points to remember are—(1) to give only a small quantity at once; (2) to pass it well back to the root of the tongue. This may often seem

a bold plan, but, in reality, it is much safer than the
more timid practice of only just passing it between
the lips.   In the former case, it immediately calls
into reflex action the muscles of the pharynx, and is
at once swallowed ; in the latter, it remains gurgling
and accumulating in the mouth, until it is perhaps
suddenly drawn into the larynx by a deep inspiration,
giving rise at once to alarming symptoms.   Sometimes it is possible. if the patient keeps his teeth
firmly closed, to pass one finger between the teeth
and cheek, and draw the cheek outwards, thus
forming a pouch, into which the nourishment may be
poured ; then, by withdrawing the finger, and keeping
the patient's head low, the liquid may often be
pressed into the centre and back part of the mouth,
when it is immediately swallowed.   This plan sometimes answers admirably, but at other times, for some
unknown reason, it entirely fails.   It is more likely to
succeed if some of the back teeth are deficient.

Fainting.—In the case of a patient fainting, the
nurse must immediately remove all pillows, and be
careful to keep the head low, even for some time after
revival.   In severe cases it may even be necessary to
support the head over the side of the bed.

Eau de Cologne, or dilute spirit, applied to the forehead, at the back of the ears, and along the hair-
parting is useful in reviving a fainting patient ; the
effect is attained by gentle blowing on the moistened
part.   When a patient is faint do not forget to
provide fresh air by opening doors and windows.   The

medical attendant should always be informed in these cases ; in his absence, if the fainting appear serious, apply friction to the limbs, from below upwards, to send the blood to the heart and brain. A piece of ice introduced in the lower bowel is a safe and often very efficient means of revival.

Fits of different kinds must often be brought under the nurse's notice. Here she must be careful not to do too much : the common practice is to forcibly hold down the patient, giving rise to an unseemly wrestling between him and several other persons. If the patient fall on the floor, the nurse should immediately place a pillow under his head, loosen his clothes, especially about the neck, and then stand by, only taking precautions to prevent him from injuring himself. It must be remembered also that some patients are subject to an outbreak of maniacal violence after a fit, and the nurse should always have assistance ready at hand in case of emergency. It is needless to say, such patients should never be employed in any work in which they are likely to injure themselves on the sudden advent of a fit. Of course, we refer here to epileptic fits ; the treatment in cases of apoplexy or hysteria needs no comment in this work.

Frost-bite.—If a patient is admitted suffering from the effects of frost-bite, the nurse must not too hastily proceed to apply warmth. Friction with snow in a cold room is at first only admissible, and any further change must be very gradual.

Foreign Bodies in Ear, etc.—The nurse should

never interfere in any case of a foreign body in the nose or ear, however easy it may seem to remove it, but should at once send word to the surgeon, or she is sure on some occasion to succeed only in rendering its removal more difficult.

Dressing Burns.—The first dressing of burns, on admission, will necessarily devolve upon the nurse, as, if left uncovered, the pain is intense. The object is to protect them from the air, and there are number-less dressings for this purpose. We select three from the list. The most common plan is to cover them with rags dipped in "carron oil," a compound of olive oil and lime water, but this is a dirty and uncomfort-able method. A better one is to cover them with flour, and then apply a layer of cotton wool and a bandage. But the best, cleanest, and most comfortable, though at the same time the most expensive plan is as follows : Spread pretty quickly some zinc ointment on strips of lint, apply these evenly to the part, next place a layer of cotton wool, and then a bandage over the whole.

Hæmorrhage.—The most serious thing a nurse must at any moment be prepared to combat is hæmorrhage. It may be venous (dark red), in which case it will nearly invariably be connected with an ulcer and varicose veins of the leg. The patient in such a case must be placed in the recumbent posture, the leg should be well raised by means of pillows, and a pad of lint and bandage adjusted, so as to make some pres-sure more especially *below* the wound. In the case of arterial (bright red) hæmorrhage (by far the more

common), any pressure must be applied *at* or *above*
the wound. Should it be only slight, the application
of cold water or ice may suffice to stop it. Where
there is a firm substance as bone beneath, pressure by
means of a thick pad of dry lint and a bandage firmly
applied over the wound will invariably arrest it,—
bleeding from the scalp, for example, can always be
thus arrested. But, above all things, the nurse must
remember, that if she can see the bleeding point, she
may always temporarily stop the hæmorrhage, by
firmly applying her finger to the spot, till the surgeon
arrives. She should never lose her presence of mind,
but always remember this cardinal point. It will
often prove more serviceable than the knowledge of
a proper application of the tourniquet, though the
nurse must always endeavour to learn the position of
such arteries as the femoral and brachial, and how to
apply the tourniquet to them. In hæmorrhage from
the nose, prop up the patient well in bed, do not let
him hang his head over a basin, and apply cold to the
forehead and nape of the neck. In cases of spitting
and vomiting of blood, raise the patient, enjoin perfect
quiet, soothe and reassure him, and let him suck some
ice, or give him only things perfectly cold.

Injuries to Head.—If a patient is brought in suffering from some injury to his head, and in an uncon-
scious state, get him to bed at once, place blankets
and a hot-water bottle to his feet, raise his head somewhat, and apply cold rags to it. Beware of rashly
giving stimulants.

The operation-room must always be kept in such a condition as to be quickly rendered ready for use in case of emergency. It must be kept scrupulously clean, with needles threaded, ligatures cut to proper lengths, sponges of various sizes in bowls, etc. When required it will then only be necessary to light the fire, get a supply of hot and cold water and ice, lay out the instruments likely to be required, and have at hand a little wine and brandy. During the operation, the nurse must be constantly ready to hand anything that may be required,—instruments, ligatures, towels, a basin in case of sickness, etc.,—she must give a helping hand wherever it may be wanted, and never be without a supply of sponges. After the operation, she will take the instructions of the surgeon as to the further treatment of each case, but will always sit by the bedside till the effects of the chloroform have passed off, as the patient may be seized with sickness, or may become restless and disturb his dressings. A sharp look-out must also be kept for any hæmorrhage.

Sponges must on no account be used for washing wounds, nor for wiping away discharges: pieces of tow, lint, or rag must be substituted, and afterwards immediately burnt.

CHAPTER VI.

SPECIAL FEATURES IN THE WORKING OF COTTAGE HOSPITALS.

Visits of patients' friends—Religious services—Office of chaplain—Management—Letters of recommendation—Medical practitioners not belonging to hospital staff—Inadmissible cases—Patients boarded by the nurse or matron—Practice at Fowey—Furnishing and fitting up a Cottage Hospital—Out-patients—Consultation by medical staff of hospital on cases sent by medical men for that purpose—Hospital kitchen—Provident and other dispensaries—Cleansing and disinfecting at a given time—Varnished paper for walls of hospitals and nurseries.

In perusing the various reports, sundry special features of interest, which are peculiar to one or two hospitals here and there, seem to demand a passing notice. We have therefore decided to devote a short chapter to their consideration, believing that the best school of practical instruction is that in which one is able to learn what to receive, and what to avoid, from a study of the experiences of others in the same path of duty or labour. We shall arrange each subject under a separate head, in order to facilitate reference, and to save time.

Admission of Friends to see the Patients in Cottage Hospitals.—One would naturally suppose that in the

(106)

little village or cottage hospital the managers would aim at making the arrangements as simple and homely as possible, and that they would encourage everything likely in any way to foster the feeling of rest and homely comfort amongst the patients, which these hospitals, as originally established, were meant to promote. It is a matter of regret that so important an element in the successful working of these institutions has often been lost sight of by those who are responsible for the rules and regulations. Thus, free permission for the friends of a patient to visit him, while in the cottage hospital, at all reasonable times, would seem one of the most desirable regulations possible.

The friends are sure to live in the immediate neighbourhood, and a daily visit, when the health of the patient permits it, would appear to be so likely to make the hospital popular, and to that extent successful, that one would naturally expect to find every facility offered in this regard. Not so, however, and the rules as to the admission of visitors require material amendment and alterations in many cottage hospitals. It is decidedly unnecessary to limit the visits of patients' friends to two hours a day for three days in the week. Such a rule is not only reasonable but necessary in a general hospital, where the work to be got through in the course of the day is very great. But in a cottage hospital, we hold that all restrictions of this kind should be abolished. The proper rules are, in effect, that "patients may be visited by their friends between the hours of 2 and 5 o'clock daily, and on special

occasions at any hour. These visits will at all times
be subject to the discretion of the medical officer in
charge of the case." "Not more than two visitors to
each patient shall be admitted at the same time, and
all visitors must strictly comply with any rules or
orders given for the good conduct of the institution."
This may appear to be a small matter at first sight,
but the patients regard it as of so much importance,
that it should not be overlooked in the best managed
institutions.

It has often been said that extremes meet, and so
it is proved in the case of this rule in cottage hospitals.
At one hospital the friends of patients are allowed to
visit them "at all convenient times," but at another,
the friends of the patients are allowed to visit them
"only on Thursday between the hours of 2 and 4
o'clock." The excessive liberty at the first hospital
has not been abused in any way, and the patients and
their friends value it as an inestimable boon. We recommend for universal adoption the freedom of the
one, in preference to the harassing and unnecessary
restrictions of the other.

Religious Services and the Office of Chaplain.—As
bearing on the above rule, we think it wise to show
how fully the homely feeling of the cottage is maintained in the matter of religion. In this respect very
little is left to be desired. The nurse or lady in charge
reads prayers daily in the convalescent room, to such
as are able to get up ; and she also reads to those
patients who are confined to bed, if they desire her to

do so. On Sunday, it is the almost universal custom, for those patients who are able, to attend Divine service in the parish church, and the rector or his curate generally finds time to visit the sick in the hospital some time in the course of the day, as also during the week. So admirably has the feeling of simple piety, which animated the early promoters of this movement, descended upon their successors and followers everywhere, that sectarian bitterness is entirely unknown to the managers of these little hospitals. We regard this as a great proof of the sound common sense which characterises the great majority of the country clergy, from which body the honorary secretaries and more active managers are in the main selected. There can be no question that the day of sickness is a great time of blessing to all if rightly used, and it is of the highest importance that the utmost religious liberty and freedom should exist in every cottage hospital.

Chaplain. —We find few rules relating to religious attendance, and these simply note, that patients may be visited by any minister of religion they desire. From information kindly given to us, we find this is the case in every hospital without exception. None are so bigoted as to exclude the clergy of any denomination. In those hospitals conducted by sisterhoods, where we might perhaps imagine more narrow views would prevail, the same rule is in force. In one or two hospitals this liberality is carried to the opposite extreme, for the rules directly provide that no minister of any denomination shall be on the committee, lest it might cause jealousy amongst others. As regards a regular chaplain, practically the vicar or curate of the parish is in the habit of frequently calling at the hospital, and seeing such patients as may desire his ministrations. Very little attention, however, seems to be paid to the reading of regular prayers and Church services, a field which is too much neglected. Surely an opportunity should be afforded each convalescent to praise Almighty God at least once a day for His goodness in restoring him to health, and to commend afflicted neighbours confined to bed in the next ward to the Almighty's protecting care.

Management.—In most cases the management is under a committee, with here and there a sub-committee or working committee in addition, often called trustees. Sometimes there is a still further sub-division, and an acting manager is appointed, as at Cranleigh, who, along with the medical officer, manages most of the details of the hospital, and reports to the committee.

Letters of Recommendation.—With, we believe, very few exceptions, a letter of recommendation from a subscriber is always required before a patient can be received ; but accidents are admitted without such letters, though such patients are sometimes expected to get one afterwards. This is not as it should be. For though letters of recommendation are valuable for the reason that the subscriber who signs them renders himself liable for the amount to be paid per week for the patient's maintenance, yet the object of hospitals is to relieve suffering, and not to make money, and too much stress is laid on getting a large and increasing balance at the end of every year.   A sum of £50 in hand for emergencies is amply sufficient for a cottage hospital.

Neighbouring Medical Practitioners not belonging to the Permanent Staff.—In almost all cases the medical men of the district are invited to follow up the treatment of their own cases in conjunction with the medical officers of the hospital.

Inadmissible Cases.—Into most cottage hospitals, infectious, incurable, and phthisical cases are not admitted.

Patients boarded by the Nurse or Matron.—At Milton Abbas Cottage Hospital, founded by the Baroness Hambro', the nurse is paid £25 a year, with an addition of 10s. a week as board wages. She is allowed a servant to assist her in the work, who is paid wages, and has in addition 8s. per week allowed her for her board. Here the patients are supplied by the committee with wines and other stimulants only ; the Baroness Hambro' gives the necessary vegetables and much of the milk free of charge. The nurse has the right to supply the patients with other food and necessaries, except, of course, drugs, etc., and she is allowed by the committee 7s. a head per week for adults, and 5s. for children. The sum received by the nurse for the year 1875, for the board of fifteen patients, etc., under this rule, was £73 10s. It cannot be doubted that such a system as this should be discountenanced on every ground of good management, efficiency, and economy.

At Fowey the arrangements for boarding the patients in vogue when the second edition of this book was published were most unsatisfactory and unbusinesslike. About five years ago the whole affair changed hands and was completely reorganised. In order, however, to show how things ought not to be done, we think it well to reproduce what was given in the second edition of this book. A regular nurse was employed, who resided in the cottage rent free, but provided her own furniture. She was paid when her services were required to attend to any sick person in the hospital; but her wages were settled by the superintendent as circumstances might require. Many patients were fed from the nurse's table, or they might provide themselves in some other way. Others were placed in the institution and daily supplied with dinner, and perhaps breakfast, by the person who got the patient admitted. Anything worse than this it was almost impossible to imagine, as we must always regard diet of proper quantity and quality, served at regular hours, as one of the most important features— nay, an essential feature—of successful hospital administration.

One of the medical officers, Mr. A. Percy Davis, commenting on the observations on this subject in the first edition of this work, remarks: "Fowey is a small seaport, the town itself containing almost 1500 inhabitants. The patients are from the poorer classes

of the town and neighbourhood, and from sailors
landed from the shipping in the harbour. The
number of patients admitted in the last two years
amounted to 40, and the subscriptions received from
the public did not exceed £15 in either year. With
reference to the dietary of the patients, I must tell
you that many of them (including all foreign and
almost all British seamen) pay for their own food.
They are all of a class that have been accustomed to
plain and homely fare, and I am satisfied that what
they get is sufficient, both in quantity and quality,
and that it is properly served. The meals furnished
by the nurse are amply sufficient in the majority of
cases, and they are supplied at a far less cost than
would be the case if a separate dietary were insisted
on. Whenever food of a superior kind is required,
it is usually provided either by the patients themselves
or their friends, or through the kindness of persons
in the neighbourhood, and sometimes out of the funds
of the hospital. By dint of careful management and
strict economy, and with the aid of an excellent nurse,
much good has been done at this place through the
instrumentality of the hospital, at a very small cost
to the public (apart from the cost of its erection,
etc.) ; and thus the institution has, in my opinion,
fulfilled the objects for which cottage hospitals are
needed." Notwithstanding this expression of opinion,
we retain our view as to the extreme undesirability
of any such arrangement as that which formerly
existed at Fowey, and we are pleased to find that

the hospital is now doing good work under the new
régime.

Furnishing and Fitting up a Cottage Hospital.—No
doubt, many more hospitals would be started, if the
would-be originators could see their way to fit up the
cottage and put it into fair working order for a start.
An excellent idea originated at Boston, Lincolnshire,
where four families each furnished a ward at their own
expense. This is a very good plan, as it not only
enables those who are not rich to give of their
collected earnings a sum which will enable them to
see the result of their contribution in an appreciable
and permanent form, but it assures to the hospital
a sustained interest in its prosperity and welfare,
which cannot but be beneficial to its continued success. We have much pleasure in drawing public
attention to this specially admirable plan, in the
belief that it will encourage other persons, whose
individual means may not be large, to go and do
likewise.

Out-Patients.—It seems to be the usual custom to
give eight out-patient tickets for a guinea; but at Savernake only two out-patient tickets are given for this sum;
and the principle of small subscriptions is introduced
at Beccles, where 6s. is the price fixed for one out-
patient ticket and 10s. 6d. for two, and at Newton,
where the subscription for one out-patient ticket is 5s.
and for two, 7s. 6d. It is the invariable rule not to admit
as out-patients those who are able to pay for their treatment, paupers, or persons in receipt of parish relief, and

members of clubs or sick societies. Every out-patient
is bound to bring a subscriber's recommendation, or the
case is refused treatment. The decision as to the circumstances of the patients, and their ability to pay,
seems to be left in these cases entirely to the discretion of the medical officer. Under no circumstances
are patients allowed to be visited at their own homes,
as under the dispensary system, from any of the cottage
hospitals mentioned above, and unless an out-patient
can attend at the hospital, his case is deemed in-
eligible, and treatment is refused. The time for which
a ticket lasts varies from six weeks to three months.
Out-patients are required to bring their own bottles
and gallipots, etc. At Crewkerne the following rule
is added, and we cannot help thinking that it is likely
to be of value, and to lead to amicable relations between
all the members of the medical profession and the staff
of the cottage hospital : " The medical officers will con-
sult upon the case of any patient sent to them by a properly qualified medical gentleman for the purpose on
Thursday at 10 A.M." We believe that, in cases of
difficulty, this offer will be accepted gladly by outside
practitioners.

On the general question of the advisability or otherwise of opening out-patient departments in connection
with cottage hospitals, it is difficult to give a decided
opinion on the facts before us, one way or the other.
We fear, however, that the funds of most cottage
hospitals will hardly bear this additional burden, and
that, as a rule, the spending of their income for such

a purpose decreases materially the usefulness and
efficiency of the in-patient department.

A Hospital Kitchen.—At Shedfield the hospital
kitchen is an independent branch of the institution.
" Tickets can be procured by any one (whether subscribing to the hospital or not) who desires to help
the sick and needy, by providing nourishing and well-
cooked food." They may be had throughout the
year—meat tickets, 9d. each; broth and pudding
tickets, 6d. each—from the matron at the hospital.
Each ticket will procure ½ lb. of cooked meat, and
broth or pudding as required. The tickets have to be
presented at the hospital on Mondays and Thursdays,
and the food is given out on Wednesdays and
Saturdays at the hospital at one o'clock. This is
certainly an original idea. We are informed, however,
that the kitchen no longer seems, as formerly, to supply
a demand. Offering tickets to village shops proved a
more popular form of help, and the plan has now been
suspended, though in severe winters the hospital is
used as a centre for distributing soup. We can
imagine, in winter especially, if carefully managed, the
kitchen will be a real help to the sick poor, and we
commend the plan for adoption where such a system
may be needed.

Provident and other Dispensaries.—Provident and
other dispensaries are worked with certain cottage
hospitals. However good these institutions may be
and undoubtedly are in themselves, we are of opinion
that they are best kept separate from the cottage

hospital system. The only way of working these institutions under the same management is to keep the accounts of the dispensary and hospital quite distinct. Even then the non-success of one enterprise will probably kill the other. We are convinced, as we have stated elsewhere, that it is a mistake to mix up other schemes with that of cottage hospitals. All the writers on the subject have held these views, and our experience teaches us that, on business grounds alone, such a system is not likely to work efficiently.

Cleansing and Disinfecting at a given time.—A very good rule on this point was quoted in the second edition of this book. After seven years the hospital mentioned was closed for six weeks, the wall papers were all removed, the mattresses renovated, and the bedding was overlooked. In subsequent years, the hospital was again closed for a short time to undergo the thorough cleansing in all its parts, which the medical officer considered necessary every second year. A hint as to purity is here given, which should be followed by every cottage hospital in existence.

This purity can be effectually secured, and the annual cost of cleaning can be minimised, by having all the walls painted and varnished throughout the hospital at the outset. Experience has proved that the expensive Parian cement, which was a popular rage some ten or twelve years ago, is neither impervious nor non-absorbent. Parian, except as a smooth surface to be painted and varnished, possesses few advantages, and we should strongly advocate Portland cement in preference. A trial of silicate paint leads us to feel that its merits have been too loudly sung. The absence of lead from paint is undoubtedly an advantage; but to have walls recently painted suddenly assuming the appearance of a huge cobweb, from the multiplicity and variety of the cracks in the paint, is not calculated to commend it to the adoption of economical people. Yet silicate does undoubtedly show a tendency to crack in this way, it is apt to come off in flakes, and it is a difficult article for an ordinary workman to use successfully.

Varnished Paper for Walls of Hospitals and Nurseries. Returning to the consideration of the comparative advantages of varnished paint and varnished paper for hospitals and nurseries, a prolonged trial leads us to recommend varnished paper for preference. We find that unless the varnish or the paint is everywhere applied with great care and skill, soft patches will here and there be left, to which dust and other matters will attach themselves and thus disfigure the whole surface. Indeed, the utmost care will not ensure immunity from these unsightly patches, and the best oil paint will fade and discolour in time, in spite of the varnish. On the other hand, a good paper of tasty pattern can be selected ; the manufacturer will carefully prepare it for varnishing, and, with proper usage, walls thus treated will last without stain or disfigurement for from ten to fifteen years at least.

In applying the varnish to paper or paint we have ascertained that the best white copal varnish should invariably be used. It is a trifle expensive, but to incur this slightly increased first outlay will be found the cheapest plan in the end. Care should be taken to apply only one coat of varnish to new paint or paper. At the end of the first year, when the whole has been carefully washed down, a second coat of varnish will ensure a wall-surface smooth, glossy, hard, non-absorbent and impervious, which will wear for ten years at least.

CHAPTER VII.

Cottage Fever Hospitals — Cottage Hospitals and paying patients — Payments by patients according to their means — Arguments for and against admission to Cottage Hospitals of remunerative paying patients — Sir John Simon on pay wards at St. Thomas's Hospital — Experience of Cottage Hospitals as to mixing of free and pay patients — Special facilities afforded by Cottage Hospitals for admission of remunerative paying patients — Opinions of managers and medical officers of Cottage Hospitals on the subject — Convalescent cottages — Wards at Cottage Hospitals for convalescent patients — Necessity for convalescent institutions being under a separate roof from hospitals — Convalescent homes needed by General Hospitals — Convalescent cottages for the poor of large towns — The institution at Epping — Country holidays for poor London children — Description of the efforts of Madame Batthyany and Miss Synnot in this direction — Mortuaries at Cottage Hospitals.

THIS book is not the place for argument as to the advantages which accrue to the community from the possession of a properly organised hospital for the reception and isolation of cases of infectious disease. It will be sufficient to say that in the reports of the medical officers of health in almost every place where such hospitals are provided, the most convincing testi-

(120)

mony is given of their usefulness in stamping out epidemics which would otherwise not only have caused much suffering and mortality, but imposed heavy burdens upon the rates.   Fifteen years ago the hospital provision throughout the country for infectious cases was most defective.   Since then we gratefully recognise the vast strides which have been made in the provision of adequate hospitals for the isolation and treatment of infectious disease.   These isolation hospitals are provided by the local sanitary authorities, and it is not necessary therefore to repeat the arguments which we felt constrained to advance at the time the second edition of this book was published.

A very large proportion of the more important towns and also of the rural districts have now made some provision of this kind.   To neglect or delay to provide such an effective local defence against infectious disease is a very short-sighted policy, especially in the case of urban communities or health resorts. There are some authorities, as there are some individuals, who refuse to learn from the experience of others, and prefer to postpone taking precautions until they themselves are visited by a sudden outbreak of disease which they are unable to meet, and which, therefore, will probably cripple their immediate resources, and possibly injure their trade and reputation. This policy is in the end one of penny wise and pound foolish, and this view has now been accepted as the correct one by every intelligent local authority throughout the country.   In these circumstances we do not propose to go further into the question in this book, especially as the fullest information can be obtained in another volume entitled *Helps in Sickness and to Health*, published by the Scientific Press, Limited, 428 Strand, London, W.C.

As the cottage hospital managers have been from the outset reformers of several abuses in the English system of medical relief, it is not surprising that they should be the first to ask, whether a hospital can be utilised occasionally, or at all, for the benefit and assistance of those who desire hospital treatment, but who can pay a remunerative sum for it.

Cottage hospitals have always encouraged payments from patients according to their means. The result has been wholly satisfactory, and this system is likely to be adopted ultimately by the larger medical charities. What are the reasons for and against the admission to cottage hospitals of remunerative as opposed to poor paying patients? Let us consider the objections first.

It is urged that the difficulty of arranging the fees which the medical attendant shall receive constitutes a grave, if not an insurmountable, obstacle. The answer is simple. Let the hospital managers decline to interfere between the doctor and the patient, and let the fees be a matter for their mutual arrangement. The managers will then have to provide suitable accommodation for the care of this class of patients, and to fix a sum per week which will in their opinion compensate the hospital for food, nursing, houseroom,

wear and tear, and every other reasonable require-
ment. The arrangements of cottage hospitals seem
to favour such a system. As a rule, there is no
constant or great pressure upon the beds. There are
usually one or more small wards specially adapted
to the reception of such cases. The ordinary staff
of the institution could, as a rule, easily attend to their
requirements, and their presence will exclude no one
else.

But here another objection arises,—whether these
cases will not exclude other and poorer patients by
taking up space which is properly their inheritance.
Is this so? The objectors urge that if remunerative
paying patients are admitted, they will gradually
elbow out the poor altogether. This view cannot
be sustained, because, in the first place, reflection
makes it evident that the well-to-do are not likely to
avail themselves to any great extent of the privilege
thus proposed to be extended to them. In the second
place, only a very limited number of beds, probably
one, and at the outside two, will be placed at the
disposal of the medical staff for these cases. It must
be distinctly enacted that preference will invariably
be given to the poor as opposed to the remunerative
paying patients, and that the prior claims of the
former to all the beds in the hospital will be considered
paramount. In this way the hospital will always be
open to the fullest extent for the poor of the district
in which it is situated. In the larger general hospitals,
and especially at St. Thomas's Hospital, London, it

was objected that the admission to those institutions of remunerative paying patients, because the funds were inadequate, was little better than a speculation.

Sir John Simon, F.R.S., says on this point :—" It may be taken for granted that the project would not have been entertained, except with a view of making substantial pecuniary profits, which might considerably relieve the funds of the hospital in their present depressed state ; and the main point for consideration has therefore been to see what would be the value of the proposed undertaking, considered as a commercial enterprise." Sir John Simon, after examining the data furnished by the treasurer of St. Thomas's Hospital to enable the governors to form a right conclusion on this point, declares :—" In view of these considerations, I should myself greatly hesitate as to the expediency of the undertaking. The matter presents itself to my mind as one of a purely commercial sort, well adapted for private enterprise, but utterly uncongenial with the present functions of the governors of the hospital." [1] This question of financial difficulty does not complicate the question at the cottage hospitals, all of which are adequately supported, and so we need not further discuss it here.

It is felt by some that the admission of paying patients would tend to make an invidious distinction between high-class and low-class patients, and that such a system might but too easily derange that per-

[1] *Vide Pay Hospitals and Paying Wards throughout the World*, a second edition of which will shortly be issued.

fect impartiality of administration which is among the hospital's first duties to all who become its inmates. On this point the managers of the cottage hospitals have had some experience already. In no case, so far as we have been able to ascertain, has the mixing of pay and free patients in the wards given rise to any difficulty of the kind. At the Hospital for Women, Soho, London, and throughout America, where the pay system has long been in force, no single obstacle of the kind has arisen. Experience therefore decides against this objection.

On the other hand, the cottage hospitals offer special facilities for the admission of remunerative paying patients. At every well-managed institution, any medical man, whether he is a member of the hospital staff or not, is at liberty to attend any patient who desires his services. No medical or professional difficulty will therefore arise from the introduction of the system into cottage hospitals. In the large general hospitals this last is a great difficulty, which is everywhere recognised. There can be no doubt, as we have urged on a previous page, that in adopting a system of patients' payments, the managers of cottage hospitals ought to make provision as soon as possible for the remuneration of the medical men. Up to the present time this has not been possible ; but the large savings of some cottage hospitals make the subject a pressing one. By allowing one or two beds to be given up to remunerative paying patients, this difficulty will cease to exist. The medical staff

will then possess the privilege of using in the hospital
a few beds for their patients' cases, and this may then
be fairly reckoned as a set-off for the attention they
gratuitously devote to the poorer patients. On the
whole, we believe that there is much to be urged in
favour of a full trial of this system ; and we should,
therefore, hope to see it introduced where there is any
likelihood of its proving a success. No allusion would
have been made to the subject in this place had not
several applications been made to the author for information and advice. It may be interesting and
profitable, therefore, to conclude this chapter by giving
a few of the letters we have received from medical
men, and also from laymen engaged in the management of cottage hospitals.

A layman, J. W. C., writes :—" Our president is
very averse to our taking in patients of a paying class,
inasmuch as he thinks we shall thereby depart from
the objects of the institution. At all events, he says,
and the committee fell in with the idea, should they
come in, then make them pay both the doctor and
the hospital as much as possible. This, to my mind,
amounts to the exclusion of the kind of patients con-
templated. There is this to be said, we have not yet
opened the additional beds, so that the question is
still an open one. At present we can accommodate
nine cases; and I wonder, should the committee agree
to take in more non-paying patients, what the medical
men will say. I have hinted as much to the president.
His answer is, ' Then shut up the hospital ! ' " J. W.

C.'s letter shows how necessary it is to find a wise solution to this question before it causes disagreement and difficulty. At the hospital to which he refers, a large sum of money was recently left to the committee, and they decided to enlarge the building, having barely sufficient accommodation in the existing building for the requirements of the poor of the district. Some of the committee agreed with the medical staff that it would be desirable to admit remunerative paying patients to the new wards. There can be no doubt that this conclusion under such circumstances is sound, and for the public advantage. Without saying with the president of the hospital that all such patients shall be required to pay doctor and hospital as much as possible, it cannot be doubted that they should pay to both an adequate and fully remunerative rate.

F.R.C.S., a cottage hospital surgeon of eminence, writes :—" It must be remembered that, besides the good to the hospital patients, the local surgeon is enabled 'to keep his hand in' at bad accidents and operations, and the rich resident often gets the benefit of this on an emergency. I tell my wealthy patients that, charitable considerations aside, it is worth their while to keep up the cottage hospital for this reason alone. Personally I simply would not continue to practise here without the cottage hospital ; and I can imagine nothing more miserable for a man who is fond of his profession, than to see all the bad cases taken away from under his care to be treated at a distance. I should very much like to see a ward established in connection with the hospital for remunerative paying patients. The committee have more than once discussed the advisability of doing so, but have not yet been able to see their way. Here, as I have said before, all qualified medical men in the district may attend their own patients in the cottage hospital. The admission of remunerative paying patients would be a good way of giving the medical officers some return for their gratuitous services. For example, recently a gentleman would have given me forty guineas for performing a serious operation upon him if he could have been admitted into the cottage hospital, but, as our rules did not allow of that, he went elsewhere, and I lost my fee."

He adds, on the professional aspects of this question :—" I believe our success is mainly to be attributed to the fact that the medical officers have always pulled well together. All qualified medical men resident in the town are on the staff, and those living in the villages near may follow their own cases into the hospital, provided they associate themselves with one of the former, so that he may be called in in their absence, should any emergency arise. Great harm has, I know, often been done to a cottage hospital, by the somewhat natural complaints of some jealous medical man who has been excluded from office. The exclusiveness of the general hospitals in this respect is, in my humble opinion, a blot which calls loudly for remedy. Here the wishes of the medical officers are, moreover, con-

sulted in every respect.   Each one can admit a patient
into the hospital at any time when there is room, and
they may always have one free patient, if they wish to
do so."

The testimony of F.R.C.S. is valuable and convincing.
He is an eminent surgeon of considerable capacity,
and he has a very extensive practice.   His hospital
takes high rank amongst its fellows, and his evidence
will no doubt have great weight.   The opinions of the
managers and medical officers of many other cottage
hospitals might be quoted, but as the foregoing fully
represent the views of other correspondents, we think
they will suffice.   After conferring with many of those
who are competent to give sound advice on this question, and after a thoughtful consideration of the arguments for and against the introduction of a limited
number of paying patients into cottage hospitals, we
have arrived at the conclusion that the time has come
when a trial may with advantage be given to such
a system.   It will probably be found in practice that
its adoption will confer real benefit upon patients and
practitioners.   If this should prove to be the case, its
universal adoption will not long be delayed.   The
whole question is fully discussed, and an account of the
working of the system in foreign countries is given
in *Pay Hospitals and Paying Wards throughout the
World*, to which work we must refer the reader who
desires fuller information on the subject.

A scheme, which seems to have found some favour
with the managers of cottage hospitals, is to have a

ward or a few beds set apart for the treatment of convalescent patients, and to call the hospital a convalescent home and cottage hospital. This idea appears to us to be positively ludicrous ; for surely no one would advise his friends to leave home, and to enter a general hospital for change of air, when in a state of convalescence from a long and painful illness. Yet this seems quite as reasonable an idea as to suppose that to transfer a poor man from his cottage, where he has been treated we will say for fever, to the cottage hospital, just as he is beginning to regain his strength, will be attended with satisfactory results.

It reminds us of the story of an ardent surgeon of the old school, who started a hospital in his native town with the view of cutting out the old infirmary. Whenever an idea of sanitary improvement was mooted in the town, this old enthusiast would go to the meeting, and declare that there was no need for any excitement, as his hospital was prepared to provide the necessary accommodation. Thus, in course of time, he arranged, in a hospital of 100 beds, a maternity ward, a children's ward, a Jews' ward, and a fever ward, under one and the same roof. When it was proposed to obtain contributions to establish a sanatorium for the town, some miles distant, where the sick poor could have the benefit of country air, he again expressed his surprise that the promoters had not communicated with him in the first instance, as he was prepared to erect a sanatorium in the ground at the back of his hospital—literally the

back yard—which would fully answer every useful
purpose, and would entail only a third of the outlay !
We can quite understand—no one, in fact, can be
more fully alive to the advantages of having a convalescent institution apart from, though in connection
with a hospital ; but to have the two schemes at work
under the same roof is to our mind a needless confusion of two useful objects, to their mutual loss.
Yet at Tenby, a cottage hospital, which was established " for the accommodation of persons suffering
from severe diseases or accidents," is now thrown
open, "if *vacant beds exist*," for convalescents requiring
sea air, not suffering from infectious or incurable disease.

At Hambrook, which is established " for the
accommodation of poor persons suffering from
accidents and diseases which cannot be healed at
their own homes," and which must necessarily, therefore, be severe in character, " it has been determined,
when the hospital is not full, to admit convalescent
patients from the neighbourhood or elsewhere," in the
hope that " the good attention and *fresh air* which
they would receive in the hospital might prove the
means of entire restoration of health." We admit
the good treatment and its advantages, but the " fresh
air " is, in our opinion, at least " doubtful."

At Redruth (West Cornwall Miners' Hospital),
similar rules are in force. We cannot regard this
movement with any favour, as it must at the best be
a question, even in the most favourable cases, whether
the effect of having in the same house, possibly in

the same room, a person suffering from a loathsome
or painful illness, is likely to expedite recovery in the
case of his associate and companion, the convalescent.
We are quite of Mr. Erichsen's opinion, that from a
sanitary point of view, it is a decided advantage to
mix surgical and medical cases together in the same
wards in general hospitals. There, however, both
classes of patients are really ill ; and it seems to us
to be a Quixotic line of treatment to send the
comparatively healthy convalescent, at the risk of a
relapse, to share the vitiated air of a hospital ward.
And yet this is what the promoters of the joint scheme
of convalescent home and hospital under one roof
are striving to establish throughout the country.

Convalescent homes have long been needed by the
general hospitals ; and it would be a great boon if
every hospital, especially in large towns, could have
its own convalescent institution. Thanks to the
munificence of Mr. Peter Reid, Mr. Passmore Edwards
and others, this need is rapidly being supplied. That
it is possible to combine the cottage hospital and the
convalescent home in separate buildings, but under
one management, is proved by the experience of the
sisters of the Holy Rood, who not only attend to the
domestic management and nursing of the North
Ormesby Cottage Hospital, Middlesbro', but have at
the same time established a sea-side home for sick
children. The home receives the sick and delicate
children of the poor, giving them the benefit of sea-
air, bathing, good food and nursing, and is, in every

sense of the word, an admirably conducted convalescent home.

There is, however, one side of this question which is being gradually forced upon the attention of all thoughtful philanthropists.   Dr. Andrew Wynter, in an admirable paper on village hospitals, which was published in *Good Words* in May, 1866, showed clearly that the then prevailing plan of sending the robust countryman suffering from accident or severe illness to the large town hospital, was hygienically indefensible, because the atmospheric conditions of large cities are always adverse to the recovery of unacclimatised country patients.   We fully endorse the truth of Dr. Wynter's argument, and it cannot be doubted that such a system was needlessly risky. The widespread support extended to the cottage hospital movement has now remedied this evil.   But if the removal of country patients to large hospitals in towns was calculated to interfere with the recovery of the patients, a system which provided a means of giving the sickly, ill-developed inhabitants of a large city the great benefits of an occasional visit to the country for a fortnight or three weeks should command universal sympathy and support.   We gladly welcome, therefore, a movement which has led to the opening of several convalescent cottages with this object.

This movement originated with the managers of the Bedford Institute, who opened a convalescent cottage at Epping in September, 1875, to meet a need long felt by the committees of various branches of mission work carried on by the Institute, which is conducted by the Society of Friends. These mission agencies are situated in the densely populated districts of Spitalfields, Bunhill Fields, Clerkenwell, Ratcliffe, and Bethnal Green, places which have only to be mentioned to claim for their inhabitants the sympathy of all who know the blessing of fresh air and healthy homes, or who have seen the pale and sickly people who are compelled, by circumstances over which they have no control, to live in close and crowded courts and alleys.

It was the desire of the Society of Friends to provide for these poor people and their families fresh air, good food, quiet rest, and the influences of a pure Christian home, " whenever they were recovering from serious illness or were likely to break down under the daily pressure of a harassing life." The convalescent cottage had at first six beds, but other rooms have since been added, and there is now accommodation for fourteen patients. It is the wise determination of the promoters not to greatly increase the present number of beds, because they desire to ensure home influence and care. Men, women, and children are received. The cottage is conducted on temperance or coffee tavern principles, an arrangement which hitherto has resulted in a marked success. Inmates are admitted with the recommendation of a subscriber, on payment of 7s. a week for adults and 3s. 6d. for children. Some of the patients remain from three to four weeks at the cottage, but the average stay is about a fortnight.   The original outlay was small (under £200), the rent paid is £41 per annum, and 148 patients were relieved in 1894.   The average cost of each patient is only 50s.   We commend this scheme to the attention of the charitable, who will be able to obtain any further information from the secretary at the Friends' Institute, 13 Bishopsgate Street Without, London, E.C.

Since the writer published the first and second editions of this book he has visited very many parts of the United Kingdom, and has been struck everywhere with the fact that the amount of hospital accommodation was evidently on the increase.   Is this vast increase necessary or politic or provident?   The answer must be decidedly in the negative.   All things have a proper limit, and this proper limit is being passed so far as hospital accommodation is concerned. The impulse which prompts the provision is as noble as the results are likely to prove disastrous.   What is wanted is more accommodation for convalescents, not more hospitals.   The writer has personally inspected the dwellings of the poor in all parts of London, in many of the provincial, Irish, and Scotch towns, and in some of the country districts.   In the latter, cleanliness is the strong point amongst the poor.   In the former, the inhabitants of the overcrowded districts of large towns are, in spite of the better side of their nature, degenerating more and more into habits of uncleanliness with all its attendant ills.   If Sir W.

Lawson would go with the writer to some of the poorer districts of our large towns, and if he could be made to realise the awful condition of the dwellings of the working classes in these regions, he would have to admit, however reluctantly, that Local Option, the Permissive Bill, or any legislation of like nature will not suffice to make the people sober. Drunkenness is but the effect of the conditions under which our modern civilisation compels the poorer classes to live. The writer has often been forced to admit that the attempt "to drown their sorrows in the flowing bowl" is a very natural and probably a rational sequence of the surroundings of the poor inhabitants of our large cities. The writer's note-book reveals a state of affairs very real, and almost too terrible to reflect upon with calmness. Any system which tends to raise the inhabitants of the crowded alley above the squalor, the misery, and the discomfort of their surroundings deserves the support and sympathy of every practical philanthropist. It is with a feeling of joyful thankfulness, therefore, that we shall now proceed to explain an inexpensive but excellent plan which is being adopted by the vicars of several of the larger east-end parishes of London, with the view of giving the children of their poorer parishioners a country holiday every year.

The system of convalescent cottages to which we refer, originated in America, where it has flourished to a large extent. The method adopted is very simple. Communications are opened with the squires

and clergy of country villages, and by their means the
labourers and small farmers are induced to take a
certain number of town children for a limited period
to reside in their cottages or homesteads. Mme.
Batthyany, of Eaglehurst, was the first lady to try the
system in England. Her scheme was to get a certain
number of children from the poor overcrowded parts
of London, say six, with one grown-up woman,
either a sister or a poor over-worked drudge or
seamstress, or a mother, to come to the country for a
week. By changing the visitors every week during
the summer months, 100 poor people and children
can easily be placed within reach of the pleasures
and advantages of a week in the country.

The object is to arrange that every child shall have
one week's holiday every year, that it shall thus be
placed under conditions of happiness and health which
will give it "a memory and a hope." The idea is to
give these poor children a run in the fields or by the
sea-shore, to let them live only a little better than they
do at home, to make them feel themselves free of school,
and to place them in communion with nature and the
rest of humanity. Every village should have a few
rooms devoted to this object—all the villagers can aid
the scheme. The farmer's wife can give a little food,
—milk, eggs, butter, and the like ; the rich can help
by giving money to pay for the hire of rooms or a
cottage specially devoted to the purpose, and to
defray the cost of railway and other travelling
expenses ; the young ladies can turn their needles to

account by repairing, mending, or making clothes for
the little strangers; all can give these poor little
waifs and strays a hearty welcome and a cheery
word. The children themselves will return home
better in health, with renewed strength and spirits,
taking with them little stocks of news, little treasures
which youngsters love,—shells, pebbles, a flower-pot
with a real plant in it, and a hundred odds and ends
of things that will amuse them throughout the winter,
and which big houses at present throw away as too
insignificant to be given to their own poor. Each
child will have new subjects of interest, and it will
give its playmates the history of its week of happiness
to their mutual advantage. These little ones will
look forward to their next holiday, will make plans
not all sordid, and have ideas not wholly selfish. All
will learn how to work themselves, and will see how
others labour. Being brought into contact with a new
world and with new people, they will also see that
woods and water and sky and sunshine are pleasant
realities. Returning home, they will carry with them
a current of fresh outer air back to their pent-up city
life, to the great gain of all their little circle. Such
is Mme. Batthyany's scheme, and it merits universal
sympathy and support.

Encouraged by Mme. Batthyany's success, other
ladies have tried the experiment. Miss Synnot, in
concert with the Rev. Canon Barnett, tried the plan of
boarding out children from the east end of London for
some weeks, in the cottages and farmhouses of Miss

Synnot's own neighbourhood in the country, with most encouraging results. This was, of course, a cheaper plan than Mme. Batthyany's, and might be extensively organised could the idea be once made widely known. With the hope of helping so excellent a project, fraught as it is with untold benefits for a class of poor invalids, who at present sadly need the fresh air and country restoratives thus so easily available, the writer has gladly called attention to the subject. By a little arrangement every village might have its convalescent cottage, and the writer will gladly co-operate with any one whose sympathy is aroused by the details here given.

The Rev. Miles Atkinson, who has had much to do with the holidays provided for the poor children of St. Jude's, Whitechapel, declares as the result of his experience, that the scheme is capable of easy development. He says that the average number of children sent annually to each village varies from six to a dozen. Each relay of children remained in the country from three to four weeks. The usual payment for each child was 5s. per week, in two or three villages 4s., and in some 6s., or even 8s.; but the vicar declined to pay more than an average of 5s. per week, except for the elder children. The plan worked admirably, and was the means of doing all kinds of good to both parents and children. As it had never been attempted previously, it was free from the pauperising atmosphere which often surrounds older institutions. In many cases, when the

plan was explained to the parents, they gladly defrayed
the railway fares of their children, whilst they thank-
fully accepted the holiday as a gift and not as a
charity.

Simple as is the system, it has worked admirably,
and has done more good than can well be imagined.
Not only have the children returned home wonderfully
improved in health and appearance, but their mental
faculties have also been quickened, whilst their horizon
has been enlarged by an acquaintance with new
scenes and people. In proof of this it may be mentioned, that one little girl was so changed that her
mother did not know her, and the guard of the train
refused to give her up to a *stranger !* Moreover, the
cottagers who took charge of these poor children were
brought nearer in a very real way to town life and
town poverty. We commend the system of convalescent cottages for adults and children, to the
favourable support and sympathy of all who have it
in their power to further so admirable an undertaking.
The author will gladly attend to any letters which
may be addressed to him on the subject, and he
sincerely hopes that before long every village in the
kingdom that is capable of making some provision
for their reception, will annually receive its fair quota
of visitors from the crowded and oppressive courts
and alleys of the large towns. For the information
of those who may desire to have an idea of the forms
at present in use, the following certificates and regulations are here given.

(To be returned.)

COUNTRY HOME FOR LONDON CHILDREN.

No................

Name of child.................................... ....

Age....................

Address................ ....

Occupation of parents...

Certificate to be signed by a clergyman, school-teacher,
surgeon, or other responsible person.

I recommend...

as being the child of parents too poor to pay for.......going
into the country.

Certificate to be signed by a surgeon or physician.

I have examined...................... ....

and certify that..................is free from any contagious
or infectious disease, and may be admitted without danger
to the other children.

Date ....

(To be kept by parents.)

COUNTRY HOME FOR LONDON CHILDREN.

A fortnight of country life in the village of   , near
        , will be given, free of cost, to any *poor* London
children, on the recommendation of any clergyman, surgeon,
school-teacher, or other responsible person.   But it is hoped
that no one will be sent whose parents are themselves able to
provide such holidays for their children..

Preference will be given to girls rather than to boys, and to delicate rather than to robust children, one object being to prevent illness; but a doctor's certificate of freedom from infectious complaints will be required at the time of going to

Children are taken in rotation, as soon as there is room, from 1st March to 30th October.

The railway fare from Charing Cross. Cannon Street, or London Bridge, to   Station is   . each way for a child under twelve years of age. This has to be paid by the parents.

In the second edition we devoted many pages to the subject of mortuaries and their advantages to the whole community. We went fully into the question in every aspect, and we feel that it is not necessary for us to repeat much of what we have already said. The subject is too important, however, to be omitted entirely from a book of this character. Those who mix amongst the poor know but too well the horrors that a death causes in a crowded cottage, where perhaps the whole family, six or eight in number, are compelled to find sleeping accommodation somehow in two small rooms, badly lighted, and often worse ventilated. What must be the condition of a family like this when a death occurs? They have to choose between two evils ; for, being engaged in hard out-door labour during the day, they must perforce sleep during the night. Either some of them must occupy the room, which common decency, to say nothing of a regard for health, demands should be given up exclusively to the dead, or the whole family must shift as best they can in the other room, small and unsuited though it be at the best.

A mortuary is an indispensable adjunct to an efficient and properly equipped cottage hospital, and we have now to suggest a plan which appears to us to open up a prospect to cottage hospitals of rendering great public service, whilst at the same time making no (or very trifling) demands upon the hospitals' resources. If the mortuary attached to a cottage hospital were thrown open for the reception of the dead bodies from the town or village generally, instead of from the hospital only, this would be a great step in advance. The cottage hospitals where mortuaries exist at present are very few. At Grantham, Sutton, Savernake, Leek, Ulverston, Petersfield, and Walker, the mortuary has been built specially for the purpose, and at Ulverston the mortuary, with the sanction of the hospital authorities, is used by the public. It would undoubtedly be a great improvement if all the cottage hospitals which possess mortuaries at the present time, would throw them open to public use, under proper restrictions. At Ulverston, the committee stated in their first report, that their attention had been drawn to the great inconvenience which occurs in cases of suicide or fatal accident : the bodies in cases of this kind have, under existing circumstances, to be removed to the nearest public-house to await the inquest. In large towns, during the last few years, whenever a death occurs or a sudden accident happens, and the body or bodies are removed to the general hospital of the district, the inquest, instead of being held as heretofore at the nearest public-house, takes place in

the hospital itself, a room being set apart by the
authorities of the institution for this purpose.   If
mortuaries were built on the plan we have given, the
bodies could be kept in the smaller room, where they
might be viewed by the jury, and the inquest held in
the adjoining room, which could easily be fitted up for
the purpose. Whether the larger question is considered
worthy of attention or not, it cannot be doubted that
it is the duty of every cottage hospital committee, so
soon as funds are forthcoming, to at once build a
mortuary distinct from the main building, which the
public may have the right to use.

Inasmuch as the law gives sanitary authorities
power to provide mortuaries, and, indeed, makes such
provision compulsory upon them if the Local Government Board require, it would probably not be difficult
for cottage hospital managers to induce local authorities to make an annual subscription to the hospital
funds in consideration of the use made of the mortuary by the inhabitants of the district.   Such an
arrangement is very common in the case of infectious
hospitals established by private effort; and there would
seem no legal objection to the same power of subscription being exercised in the case of mortuaries.   Indeed,
the members of sanitary authorities, however slow to
perceive the advantages of hospital accommodation for
infectious cases, will probably be found much more disposed towards assisting in the provision of a place which
will obviate the shocking and revolting scenes too often
inevitable when a death occurs in a poor family.

The economy of the arrangement will, moreover, be an important additional factor in securing a subscription ; and we would strongly recommend cottage hospital managers to make such provision at their institutions as will enable them to accommodate dead bodies brought from any part of the district to await interment.

There is, however, one objection to this plan. One of the most useful purposes of a public mortuary is to receive the bodies of persons who have died of infectious disease. These cannot, for obvious reasons, be safely received in the same chamber as those of persons who have died from non-infectious causes, and hence the Local Government Board advise the provision of two chambers in public mortuaries provided by sanitary authorities.[1]   It would not be permissible to bring contagious cases to the mortuary of a cottage hospital where such cases, with the exception of perhaps typhoid, are expressly excluded from treatment. These facts point to the necessity for making it compulsory upon all sanitary authorities to provide adequate mortuaries for the districts they control.

The provision of mortuary accommodation is now compulsory as regards each sanitary authority in London, but outside the metropolis the duty is still optional. And yet where a mortuary has not been

---

[1] See *Model Bye-laws issued by the Local Government Board for the Use of Sanitary Authorities—XV. Mortuaries* (London : Messrs. Eyre & Spottiswoode, price 2d.). for plan and remarks.

provided in a district, a very important and valuable section of the Public Health Act, 1875, is left inoperative. Section 142 of that Act provides that "where the body of one who has died of any infectious disease is retained in a room in which persons live or sleep, or any dead body which is in such a state as to endanger the health of the inmates of the same house or room is retained in such house or room, any justice may, on a certificate signed by a legally qualified medical practitioner, order the body to be removed, at the cost of the local authority, to any mortuary provided by such authority, and direct the same to be buried within a time to be limited in such order." This section, it will be observed, is of very wide application; for it cannot be doubted that many dead bodies, not necessarily those of persons dying of an infectious disease, are in such a state as to endanger the health of the living inmates of the house or room in which the bodies are retained. Yet, unless a mortuary is provided, there is no legal power to deal with the body, and it may remain a source of the gravest injury to the public health of the neighbourhood. The establishment of a mortuary is, then, an imperative duty for local authorities; and we cannot but think that the plan we have suggested would be found at once economical and efficient.

Certainly no new cottage hospital should be built without a proper mortuary being included in the plans. Such mortuary should be built in the hospital grounds, but out of sight of the wards and

the patients. It is not difficult to make arrangements for the erection of one of these useful buildings,
without in any way exposing the patients of the
hospital with which it may be connected to any
annoyance or disagreeableness. It should have a
separate entrance, which, together with the road leading to it, may be hidden by trees and shrubs, or built
out of view from any part of the grounds.

When fitted up with every requisite and made fully
capable of meeting all ordinary requirements, we
estimate the cost of a building, according to the plan
recommended on another page, at £200 or £300, varying
with the extra outlay which a good distance from the
main buildings will entail, if it be walled off with a separate approach, as is important if it be intended for the
benefit of the inhabitants generally. Dr. Hardwicke,
the late able and energetic coroner for Central Middlesex, was of opinion that a small brick building with
two compartments, suitable for a village, need not
exceed the sum we have named, including mortuary,
two waiting-rooms, and a disinfecting chamber. This
was the cost of the one in the large London parish
of St. Luke's, and we know of no reason why, with
proper economy, it need be exceeded.

CHAPTER VIII.

COTTAGE HOSPITAL CONSTRUCTION AND SANITARY ARRANGEMENTS.

Size—Considerations to be had in view—Site—Construction—
Number of wards—Height—Plan—Preparation of walls,
etc.—Day-room for convalescents—Operating-room—Mortuary—Laundry—Ventilation and warming—Water supply
—Disposal of excreta—Earth- and water-closets—Drainage—Disposal of sewage—Admission of enteric fever cases
into wards—Urinals—Disposal of slops and kitchen refuse.

IT is not intended in this chapter to enter fully into
all the apparatus necessary for the proper sanitary
arrangements of a cottage hospital. Such an undertaking would require a volume of itself. We shall
only aim at placing before the reader the different
plans that may be useful in different cases, pointing
out the ideal towards which he should work and the
faults to be avoided, while the details themselves must
be left in the hands of the architect and builder.

Every case also must be treated on its own merits.
The circumstances in which it may be advisable
to erect a cottage hospital are so various, that no
special rules can be laid down for universal adoption.
We must be content, therefore, with a review of the
different questions that may present themselves, and

(148)

with pointing out means by which any obstacles may be surmounted.

The first question that arises is as to the size of the hospital. Hitherto it seems to us that this question has been treated on much too limited a scale. In most works on the subject it is stated that this class of hospital, to act essentially up to its character, must not accommodate more than six patients. Practically, the question assumes a different aspect; for we find cottage hospitals now in existence with beds for fifteen, twenty, and even forty patients. It must be admitted that a hospital to accommodate forty patients can scarcely be called a cottage hospital in the true acceptation of that term, and it will perhaps be convenient to limit the term to a hospital of from three to twenty or twenty-five beds, according to the requirements of the district. The smaller sized hospitals, of from three to six or eight beds, will be found, as a rule, more suitable for agricultural districts, where the villages are small or scattered, and amongst the population of which accidents are comparatively rare. But when we come to the large mining districts, the state of affairs will be very different, and such a hospital would be of scarcely any assistance. There, village has grown upon village in quick succession, to keep pace with the growth of works, and in almost all such cases there is dense overcrowding of the population The villages are often at a great distance from the county hospital, severe accidents are of frequent occurrence, and a

considerable proportion of the inhabitants is composed of young men from a distance, who occupy the
position simply of lodgers, and who, when prostrated
by severe illness, are dependent on the variable charity
of those with whom they reside.   To this latter class,
especially, the establishment of a cottage hospital is
an inestimable boon.

What considerations must guide us as to the size
of hospital required for a given district?   Most
authors on the subject try to lay down a definite
rule, based on the ratio of beds to population, but
the conclusions at which they arrive are so various
as to be practically worthless.   Thus one author
advises 5 beds for each 1000 inhabitants in an
agricultural district, whilst others estimate the ratio
as low as 1 bed for each 1000 population.   But it
it is a mistake to look only at the number of inhabitants.   We ought also to take into consideration the
vicinity of large works, mines, or factories, the nature
of the works, the ventilation and quality of the air
in mines (often very varying), the usual number (or
perhaps rather the greatest number) of accidents
under treatment at any one time, and the liability
of those employed to acute diseases from the nature
of their work.   Moreover, should it be decided to take
in cases of enteric fever (as for reasons to be presently
stated it might be advisable), we ought also to
consider some of the sanitary aspects of the neighbourhood, especially the water supply and the means
for the disposal of excreta, both of which in most

villages will be found to be simply abominable, while cases of enteric fever always prevail.

From experience gained in a large mining district the proportion of 3 to 4 beds for each 1000 *workmen or miners*, and then a calculation for their wives and persons employed in other ways, at the rate of 1 bed for each 1000 persons, does not seem at all too high. In an agricultural district the calculation of 1 bed to each 1000 persons will probably be sufficient.[1]

As regards nursing arrangements also, we shall find a hospital of 20 to 25 beds to be a very convenient size. Each ward of 10 beds would thus require the services of one day-nurse, who would also have charge of the single-bedded ward on the same side, while the night duty could easily be undertaken by one night-nurse. This would be much more convenient and satisfactory than several small wards of 2 to 3 beds, as in the smaller hospitals of perhaps 8 beds, with, as is often the case, no person on duty at night, and only a sort of combined nurse and cook in charge during the day. Yet such is the arrangement that admirers

---

[1] A surgeon to some extensive iron works in the North writes:—" I have mines here in which about 1500 men are employed ; there are numerous other mines round about. Four bad accidents at a time are not at all a rarity, and they often come in runs, besides which there must be an allowance of 1 bed per 1000 at least for medical cases. At one mine, during the last month, employing only 200 men, there have been a case of depressed fracture of skull—one of fractured thigh— also a badly comminuted fracture of humerus—all which would have been more advantageously treated in a cottage hospital."

of the cottage system pure and simple have advocated.
To our mind this would be nothing but a hopeless
conglomeration of nursing and cooking, the nursing
being probably in charge of a convalescent while the
cooking proceeds, or the cook, in the midst of her work,
being liable to be suddenly called upon to attend to
an open wound.   Evils enough, to say nothing of the
disagreeable idea to an invalid of having the constant
smell of cooking before him, which would certainly
destroy the little appetite he might possess.   In our
opinion this attempt to establish at all hazards the
homeliness of the cottage, as a substitute for the order
and regularity of a hospital, would be a very dear purchase.   Besides, this homeliness may be attained by
other means.   The numberless essential details connected with the proper working of a hospital cannot
be carried out with due benefit to the patients without
the observance of the strictest order and method.   Let
us inquire what is the peculiar charm of home life to
these people.   Every one having any acquaintance
with the cottage life of the miner class, will understand
that its charm consists in its independence,—that is,
in each member of the household doing what seems
right in his own eyes, and acting separately from the
others.   It is certain that the most extreme advocate
of the cottage system would not advise the introduction of these customs into a hospital.   Nor do we
find the labouring classes themselves objecting to the
appearance and order of a hospital, though they certainly
do object to removal to the county infirmary, where

they would get the best medical advice. We must therefore look further for the charms to them of a cottage hospital, and shall always find these to consist in the hospital being placed near at hand in the midst of their friends and relatives, without such restrictions as to the visits of their friends as may be necessary in a larger hospital; in the nurses being better known to them, at any rate by repute, than would be the case in a large hospital; and in their being allowed to have the attendance of their own medical man, who has probably attended them for years. These are home advantages which would counterbalance the appearance of the wards; and we may dismiss this strange bugbear of many minds, whilst at the same time we apply all the principles of hospital construction and good order (that have been worked out so much of late years) to the erection of our building. By these means we shall ensure much better hygienic arrangements than would be possible if we listened to the advocates of the so-called simple system.[1]

Site.—The rules which follow will apply to most other hospitals, *if we are able to select our site.* (1)

---

[1] The surgeon quoted before writes on this point:—" About here I generally find there are no regular meal times, the different members of the family rising, and coming in for meals, as suits best each individual member. So, too, there is no regular place for things—on coming in, boots and clothes are taken off and thrown anywhere—method and order are detested. It will be objected that this ought not to be the case in a properly-managed cottage, but I find it is the case in 90 out of 100, and these are just the people who would object to hospitals."

It should be dry, on a deep bed of sand or gravel
if possible, otherwise well drained.   (2) The aspect
should be south or south-west, and protected from the
north and east.[1]   (3) It should not be too near large
buildings, nor so surrounded by trees as to interfere
with the free movement of the air.   (4) It should not
be on low marshy ground, but, if possible, on sloping
ground.   (5) A site where the earth has been exca-
vated and the space filled up with débris and filth
should be avoided ; but where there is no alternative
but to utilise a site of this kind, the rubbish must be
carefully removed and the space filled in with hard dry

[1] It is of course the aspect of the wards that claims the
greatest attention. The best is north-west and south-east, the
maximum of sunlight without excess of heat being thus obtained,
the windows on one side getting the morning sun, the mid-day
or hottest sun striking the ward obliquely, and the windows on
the opposite side getting the afternoon sun. We have seen
plans thus :—

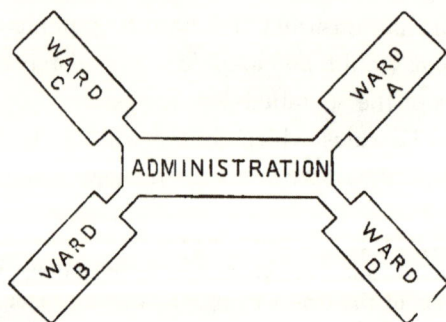

In this plan, if A and B wards are best as regards aspect, C
and D must evidently be worst.

rubbish. The surface must then be covered with a layer of cement concrete finished with a surface of asphalte or of cement and sand. The provision of a layer of concrete with an impervious surface beneath the whole area of the building is a necessity whatever the nature of the sub-soil may be,—except, perhaps, where the latter consists of granite or other equally impermeable stone,—it being of the utmost importance to prevent the passage of either air or moisture from the sub-soil into the interior of the building. (6) We should try to ensure a good supply of water. This will often be a very difficult matter, and no light consideration in any locality. (7) We should try to get an easy means of excrement disposal. (8) The hospital should be placed in the centre of the district from which its patients will be drawn. (9) It should be near the residence of some medical man who could take chief medical charge, and who could easily be called in cases of emergency. (10) We must also try to place it in its own grounds, which should be cheerfully laid out for the exercise of convalescents.

Construction.—It will be convenient to divide this branch of the subject into two heads—(1) construction of the smaller hospitals of less than 10 beds ; (2) construction of the larger hospitals of 20 to 25 beds.[1]

---

[1] (1) Petersfield (see plan) may be taken as an example of the first class of hospital. (2) Reigate (as shown on the plan) for the second.

The first class of hospitals will probably be situated in agricultural districts, with pure, good, and wholesome air, where few serious accidents occur. The cottage character may be strictly carried out in building a new hospital, and it is always advisable that the building should be new, if funds will permit. Or, if a suitable farmhouse or cottage can be obtained, it may be made available for the purpose, provided it be entirely refitted, according to the plans hereafter to be discussed under their different headings. From a consideration of the arrangements existing at cottage hospitals of this class, we propose the following plan, which, modified according to circumstances, will generally be found a suitable one :—

Ground Floor.—Kitchen, scullery, general sitting-room, store-room, small dispensary, operation room, closets and urinals for staff.

First Floor.—Male ward (three to four beds), female ward (two to three beds), single-bedded ward, nurses' sleeping rooms, bath-room, closet, etc.

In the basement, there must be a cellar for wood and coal, beer, wines, etc., and there should also be a small detached mortuary, with conveniences for making post-mortem examinations.

In a new building the closets, urinals, and bath-room must be placed within easy access of the wards, from which, however, they should be carefully isolated by a cross-ventilated lobby or some other arrangement. If it be decided to transform an old farmhouse, it will be found in almost all cases that the only closet consists of a cesspit situated at the bottom of the garden. This must be thoroughly and carefully emptied, disinfected, and filled up, and fresh arrangements substituted. The best and cheapest plan will usually be to build out a tower from the wards, with which it may communicate by a short cross-ventilated passage. Upstairs in the tower may be placed the bath-room and closets, while on the ground floor will be the scullery and nurses' closets.

We would also urge the propriety of isolating the kitchen, if possible, from the wards. This will not perhaps be practicable in cases where a farmhouse or other building is altered for the purpose ; but the point should be borne in mind if a new hospital is to be erected, for nothing is so nauseous to a sick person as the constant smell of cooking.

In any case, more attention than is usual should be paid to the ventilation of the kitchen and kitchener. It will be observed that in our model plan there are no rooms over the kitchen, so that it may be ventilated in the roof. Mr. Lanyon of Belfast has patented an excellent apparatus for ventilating ranges and kitcheners, which by means of a canopy, rare-fying chambers, and gratings, carries off the steam and smell of cooking.

In an old building, also, we shall be almost sure to find some of the walls damp and mouldy, due either to the attraction of water from the ground upwards, or to the use of soft, porous bricks. In the first case, we ought to construct a well-drained dry area round the bottom of the house, and put in a proper course of damp-proof bricks or slate. In the latter case, we might use some of the patent compositions prepared for the purpose.

Next, as to the construction of the larger sized hospital, taking as an example one with about twenty beds.

The first question that arises is the size of the wards. Shall there be two wards, each containing about ten beds, or shall there be several smaller wards of four or five beds each? Let us first insist that, in every hospital of this size, there must be at least one (perhaps two) single-bedded wards for isolating doubtful cases, or for treating serious cases of operation. Small wards for three to five patients each are to be found at Lytham and Bromley, but the only argument in their favour is that they carry out the idea of the home cottage hospital system,—as we think, to excess. They are certainly not so conveniently arranged for nursing, ventilation, supervision, and other hospital and administrative purposes. We have assumed that a hospital of this size will only be required in a place where bad accidents are of frequent occurrence. Moreover, in the cottage hospital there will mainly be cases of acute disease, while in the county hospitals many chronic cases are mixed with the acute ones. Here, then, all the arguments for the proper ventilation of a hospital press with double force, and all the latest improvements for carrying them into effect should be utilised. Now, what are these improvements? Applied to this class of hospitals, we may probably say that there are two essential objects to be kept in view—(1) the nursing; (2) the ventilation. The nursing should be so arranged that one nurse may easily and constantly overlook all the beds under her charge. If we allot a ward of ten beds, and a single-bedded ward, to a nurse on each side, we shall provide sufficient nursing for one person to thoroughly carry out. As regards ventilation, all authorities agree now that the most useful plan is the pavilion system; and some modification of this, according to circumstances, must be advised.

Another consideration that arises is the height of the hospital,—shall it be of one or two stories? Where practicable, a central building for administration of two stories, with wards of only a single story, must be advised,—(1) for conveniently separating the sexes and nurses, so that one does not interfere with the other; (2) for purposes of ventilation; (3) for increased roof space, thus ensuring a larger supply of rain water. It must be admitted, however, that two wards, one above the other, will be much cheaper, and as a rule will be preferred, especially as the single story may entail the cost of an extra nurse. Nor do we think the arguments for a single story are so forcible as to make us insist on this plan being absolutely essential to perfect sanitary requirements. We shall probably be told that the erection of a hospital of this class on the pavilion plan is quite impracticable on account of the expense. If, however, the advantages of such a plan are so great, let us consider whether this objection will not disappear. This class of hospital will be required mostly in those districts where there are large works or mines, the owners of which are, as a rule, wealthy, and only require to be shown the necessity of the undertaking to contribute generously to its success. We must bear in mind, also, that all fatal mining accidents are carefully investigated by the Government inspectors, and every mining proprietor is exceedingly anxious to contribute in any way to the successful issue of these accidents. Many of them, also, will give in kind as well as money. Thus one, perhaps, will give the land, another the bricks, and a third will find the labour. By these means a hospital may often be erected at comparatively little cost, and we venture to think that the cases are few indeed where this plan will not be found practicable, if only it is sufficiently urged.

Let us now map out a plan of the rooms that will be required in our miniature pavilion hospital.[1]

Central Administrative Block (Two Stories).

Cellars for wood and coal, beer, wine, etc.

Ground Floor.—Kitchen, scullery, matron's or nurses'

---

[1] The Ross Memorial Hospital, though only for eight beds, is a creditable example of the pavilion system. The wards contain only two beds each; otherwise, with larger wards, it would strictly answer this description. It cost £2500 to erect, and no expense was spared in any way. The furniture and fittings entailed an extra expenditure of £300, and at the present time it may be regarded as a very complete little hospital.

sitting-room, mess-room, store-room, dispensary, opera-tion room, closets, urinals, etc.

First Floor.—Bedrooms for matron, cook, night nurses, under nurse—(staff).

Cross-ventilated lobby leading from central block on each side if wards are of a single story, or from one side or centre, with staircase, if of two stories.

Wards.

Male ward (ten to twelve beds). Female ward (eight to ten beds). Two single-bedded wards.

It will be useful now to take the rooms separately as far as necessary, and note the chief requirements of each. The wards will have a nurse's room at one side of the entrance, and a scullery on the other side, while at the opposite end will be the bath-room and closets. The walls must be of some non-absorbent material. The best Portland cement, or some good strong ordinary plaster with a smooth finish, should be used, Parian cement being expensive, absorbent, and unsuitable. It is a very good plan to paint the walls as soon as dry (four coats), and afterwards to give them two coats of best copal varnish. This will be more expensive at first, but the primary outlay will be repaid over and over again ; for when this plan is thoroughly carried out, a perfectly smooth, hard, impervious and non-absorbent surface is presented, which can easily be washed down, and the wards are thus capable of being readily and completely disinfected. Where this plan has been tried pyæmia has disappeared, and the most satisfactory results have been obtained. Walls thus prepared will remain perfectly clean in appearance, being at the same time very generally safe for from ten to twenty years. The ceiling should be lime-washed, a process which must be renewed at least once a year. In transforming an old farmhouse or other building, the ceilings and walls must be first well scraped, and then one of these methods applied. In a new building great attention should be paid to the floors, which are best made of oak with perfectly tongued joints. The boards should be " side " or " secret " nailed, so that no nail-holes appear on the surface, and when laid should be either treated by the paraffin process or wax-polished. The paraffin process consists in forcing, by means of irons made for the purpose, melted paraffin wax into the pores of the wood ; the surface is then cleaned off and polished with a preparation of liquid paraffin. It is claimed for this process that it not only gives a perfectly impervious and antiseptic surface, but that it conduces to the preservation of the wood. The great advantage of both this and the wax-polish is that the floor can be cleaned by dry rubbing ; and the evaporation which goes on after scrubbing in the ordinary way, a process so fraught with danger to the patients, is entirely avoided.

If of a single story, the flooring must be raised above the ground and the space below well ventilated.[1]

[1] Dr. Swete advises that it should be raised on arches with a platform covered in like a railway platform, to form a terrace walk for convalescents. This seems a good idea.

As before explained, a layer of concrete or asphalte should extend over the whole area of the building. The finishing of the interior should be as plain as possible—cornices and other embellishments only afford lodgment for dust and dirt—and all corners should be rounded.

A day-room for convalescents will be a most useful adjunct, and this should be finished off and fitted more in the cottage style, thus realising the object which most writers think so important : for here both sexes might take their meals, work, read, and amuse themselves. The walls should be hung with cheerful pictures, and books and book-shelves provided. Outside the day-room could be erected an open verandah, in which the patients might take exercise during bad weather. Surely this plan of thus separating the convalescents from the sick, and giving them recreation and routine as in their own homes, is more to be commended, on the score both of hygiene and comfort, than the cottage system indiscriminately applied to the wards. The beer and wine must be kept in a separate cellar from the wood and coal. There should also be close at hand a light place from which it may be given out, while the key should always be in charge of some responsible person. The kitchen should be fitted with a window or wooden frame to open at will, whence the diets may be distributed and inquiries answered without persons continually intruding into the kitchen itself. Some source of hot-water supply in connection with the grate will also be needed, and, if necessary, an earth-drying arrangement for the earth-closet. Convenient cupboards, shelves, hooks, etc., for kitchen furniture and utensils are also requisite.

The scullery should be furnished with a supply of hot and cold water, sink, plate drainer, arrangements for cleaning knives, boots, etc.   It should also be within easy access of the kitchen.   The matron's sitting-room should adjoin the store-room, or be situate in its immediate vicinity.

The ground floor is the best for the operating room, but it may be placed on the first floor, if the staircase is of sufficient width to admit of the easy carriage of patients.   If placed upstairs it can be lighted from the roof by a skylight, and there should be lamps, conveniently arranged, in case of an operation at night being necessary.   There must also be an operating table, either one of those sold for such a purpose, or one with movable head-rest and slides for the feet, which could easily be made by the village carpenter. The operating table should be a shallow, water-tight wooden tray, about 2 inches deep, 5 feet 6 inches long, and 2 feet 6 inches wide, on legs of a proper height.   On this should rest an operating plank, which should be 6 feet long and about 2 feet wide. The tray and plank should be thoroughly paraffined. The shallow tray should have an opening in its lower part, with a tube, so that all the fluids can be conducted into a bucket beneath.   Such a table is as thoroughly aseptic as the more costly ones of plate glass and iron, and keeps the patient and operator dry, while,

at the same time, it permits of the free use of
irrigating fluids. The instruments can be kept
ready at hand in a proper cupboard. Should this
arrangement of placing the operation room upstairs not be deemed convenient, a room leading
out from the corridor, with a glass roof, might be
erected. In practice it is found that an operating
theatre should always be placed on the ground floor.
This can easily be arranged with a little trouble, and it
will often prevent accidents to patients in their transit
to and from the rooms. The operating theatre at
Savernake is one of the most recent.

The mortuary should be detached from the main
building, out of sight from windows of rooms occupied by patients, and should be provided with a
table, good supply of water, skylight, drawers, etc.
In one cottage hospital the mortuary is public,—that
is, for the use of the village. This is a very good plan,
and, as already explained, is highly to be commended
for general adoption.

Laundry.—A separate out-building should be used
as a wash-house, great care being taken to place all
infected linen in the disinfecting tank, tub, or trough,
which will, of course, be provided for this purpose.
Formerly experts advised that where practicable a
constant flow of water should be directed on to all the
linen in this receptacle. Modern sanitarians, however, object to a constant flow of water, as they maintain that it simply dilutes and washes away the disinfectant solution, in which the foul linen should be

allowed to steep until it is ready to be put in the vessel to be boiled.

Ventilation and Warming.—The wards must be of such a size as to allow 100 square feet of floor space and from 1300 to 1500 cubic feet of air to each bed. These conditions in a ward of ten beds will be fulfilled by the following dimensions :—length 40 feet, width 25 feet, and height from 13 to 15 feet. The same scale should be used for smaller wards, and also in allotting the number of beds for rooms transformed into wards. Since these wards will be used only for cases of acute disease, accident, or operation, we must hold it to be extremely important that this, the usual scale for hospitals, should not be lowered.[1]

For ventilation purposes a simple system will be wanted for use in the summer ; but, during the winter, arrangements must be made by which the air will be warmed before entering the wards.

The windows should be opposite each other, and should open at the top and bottom ; the top sash should be fitted with a brass slotted plate, by which it can be opened and closed with a long arm.

Slanting valves over the windows, somewhat like Sheringham valves, as in many recent hospitals, are very useful. They open inwards, thus throwing the air towards the ceiling before it mixes with that in the ward ; they should be fitted with glazed cheeks

---

[1] Dr. Swete considers that, on account of the pure state of the air in country districts, 800 to 1000 cubic feet to each bed will be sufficient. Probably 1000 would be enough to allow.

inside to prevent down-draughts, and should be opened
and closed either by means of a long arm and a spring
catch, or by one of the various screw apparatus made
for the purpose. Sheringham valves in the walls will
often be of service. In a one-storied building, Mr.
McKinnell's system by two hollow cylinders could
very easily and cheaply be applied, and the openings,
which must be closed when warm air is to be admitted,
are not so easily shut by the nurses in this as in the
former systems.

The late Mr. W. Eassie. C.E., thus described this system :—
Mr. McKinnell's system of ventilation. upon the principle
of the double current, was invented by him in 1855, and consisted of an automatic apparatus. to be fixed in the ceiling
and roof, which not only provided a
steady influx of fresh air, but discharged
the air vitiated by respiration or combustion. The apparatus consists essentially of two tubes, concentrically
arranged, and opening at their lower
ends into the apartment to be ventilated. These tubes communicate with
the outer air at different levels, the
respired air rising up the central tube
and passing off at the higher level. and
the fresh air entering the annular passages between the inner and outer tube
at a lower level, and descending into the
room below. The inner tube projects
a short distance above the outer tube.
and is capped by an ornamental cover,
from beneath the edges of which the
foul air escapes.

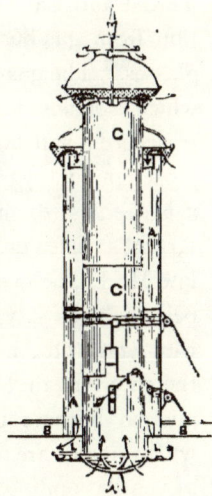

A screen of perforated material covers the escape opening,
and prevents the inroad of foreign matters into the tube. Both

of the tubes are so proportioned, that the sectional area of the centre tube is about equal to the sectional area of the annular passage comprehended between the two tubes.

Mr. McKinnell, in this invention, arranged for a partial or total closing of the inner tube by means of a kind of throttle-valve, set upon a transverse spindle inside the tube, and weighted on one side, so as to have a tendency to maintain a vertical position, and leave the passage full open. The valve was, how-ever, under control by means of a cord and pulleys. When the valve appertaining to the outer air passage was drawn up as high as possible, it completely closed the down draught passage; whilst, by letting it down more or less, the passage was corre-spondingly opened, and the current of air impinging upon the valve plate became deflected and spread out horizontally, and so became more uniformly dispersed over the room.

In transforming a farmhouse, the vertical system of ventilation would be found useful. This has been tried with satisfactory results at the Central London Throat and Ear Hospital. The products of combus-tion from gas-burners are not well carried off by bell glasses. The gas-burner should be placed in a lantern which is enclosed on all sides except the bottom in order to secure a satisfactory current of air. These are the most simple ways of ventilating without warm air ; and it is needless to enter fully into the subject of cowls, Arnott's valves, etc., as they can easily be read up in the late Mr. Eassie's very useful little work. One warning is perhaps necessary,—never use any perforated zinc, fine wire-work, etc., materials often used to finely divide the entering air, but which soon get clogged with dirt, and so become useless ; for there is no subject on which nurses are more ignorant and careless than that of ventilation.

The next aim must be to ensure a good supply of warm air to the wards in winter, at as cheap a rate as is compatible with efficiency. In this there is no need to consider other arrangements for ventilation that can possibly be blocked up, for we may be sure that to act in this way will be the supreme object with the nurse as soon as there is the least sign of cold weather, or the arrangements will only be in use during the short visit of the surgeon. Heating by hot-water pipes will be too cumbersome for hospitals of this size, and in reality the question is narrowed down to the kind of stove or fire-place most efficient for the purpose. No English labouring man's mind will be content without seeing a cheerful stove or fire-place in the ward. It is not that it gives off more heat than could be obtained by other arrangements, but in England its use is so common as to be indispensable, without totally obliterating all idea of homely arrangements. Luckily there are several very efficient stoves now in common use, by which the entering air can be warmed; and we may instance

SIR DOUGLAS GALTON'S VERTICAL STOVE.
A, plan; B, elevation; C, section.

the mode of operation in a few of them as examples of their class.

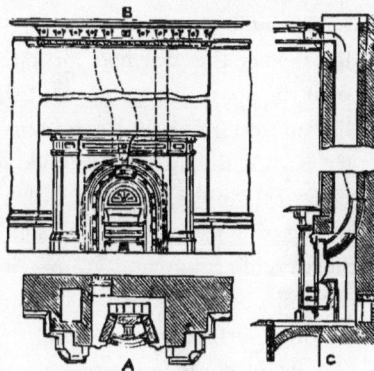

In Sir Douglas Galton's stove[1] the air is warmed
in chambers behind the grate. The point of discharge for the warmed air is below at various orifices,
while the outlets are usually provided above by foul
air shafts (with Arnott's valves) round the chimney.
Galton's is the best-known grate of a description
uniting both iron and fire-lumps. Copious supplies of
fresh air pass through the grating at the back of the
chamber, and when warmed in the latter it rises up
the hot-air flue leading out of the top of the chamber,
and is delivered into the room near the ceiling-line, or
elsewhere as may be desired.

The smoke flue has no connection with the hot-air
flue, and, if the grate has been properly fixed, nothing
can work better. It has stood endless tests, and will
perform the hardest work with a minimum of attention. Its performances have been carefully summed
up by General Morin.

The bulk of the ventilating grates of the present
day deliver the warmed air either under the mantel
or in the space formed around the grate itself. In the
Thermoson grate, which was used largely by the late
Mr. Eassie, the fire-basket, lined with fire-brick, is made
to project, and is provided with a movable iron canopy,
having gills upon it. The back of the fire-basket has
the arch of the stove cast upon it, and has also projecting gills behind. By this arrangement an increased
heating surface is secured, thus warming the air
coming in contact therewith. Apertures in the arch

[1] This stove seems to be in use in several cottage hospitals.

and front, communicating with an air chamber formed
of brickwork behind the stove, provide for the free

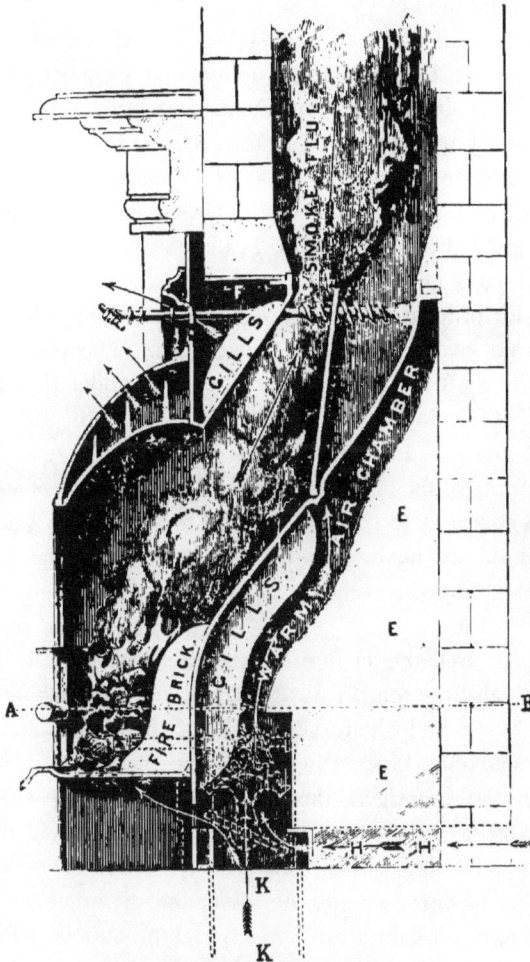

THE THERMOSON GRATE.

E, E, E, air chamber formed of brickwork; F, covering-in plate,
resting on bar G to render the joint air-tight.

circulation of heat, and for the necessary supply of

cold air. A valve is provided, by which the draught can be regulated at pleasure, either to the air chamber or to the fire.

There are several peculiarities in the Thermoson which can be sufficiently understood by a study of the large-sized section on the previous page.

A very useful kind of grate is the improved form of the excellent Pridgin Teale grate made by Messrs. Teale and Somers of Leeds. The grate, as originally designed by Mr. Pridgin Teale, consisted of a fire-brick back and sides, with an arrangement under the fire for preventing the ingress of currents of cold air and for keeping the under space hot. The effect of shutting off the cold air from the space under the fire and so forming a hot air chamber below the fuel is that more perfect combustion is obtained with greater economy of fuel. The use of fire-brick in place of iron also conduces to the same end, as fire-brick stores and accumulates heat while iron radiates it, and as Mr. Teale remarks, " chiefly in directions in which the heat is least wanted." The Teale grate has its back bent or arched over in order that the heated fire-brick may raise the temperature of the gases resulting from combustion and help them to burn, thereby lessening the smoke and increasing the radiant heat.

In the Calorigen the air is warmed by passing through a spiral tube placed within the stove, and communicating with the outer air ; there is a constant stream of warm air into the room, and the other openings are probably all converted into outlets. The

Calorigen burns very little fuel, and its price is about six guineas. We take the following from the late Mr. Eassie's admirably exhaustive Dictionary of Sanitary Appliances : [1]—

WARMED AIR FROM WITHOUT.—To stoves of a description which effect all the purposes of the Galton and other succeeding grates, it has been common of late to apply the name calorigen, or calorifère, the former word being of English and the latter of French origin.

The best examples, and indeed the only ones much in use, are the calorigens invented by Mr. Richard George, whose earliest experiments I witnessed with great interest. In 1867 Mr. George introduced the calorigen in which coal was used, and which I will now proceed to describe.

The coal calorigen consists of a chamber of thin sheet iron, the proportions of such chamber being large relatively to the dimensions of the fireplace or grate in which the fuel is burnt, in order that the products of combustion may circulate within such chamber some time before passing into the chimney or flue. The grate or fireplace

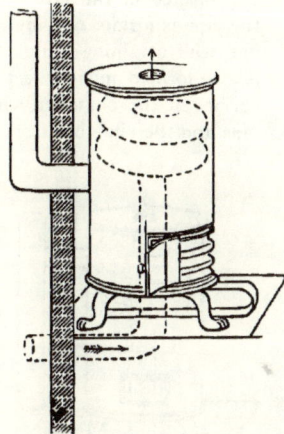

GEORGE'S COAL-BURNING CALORIGEN.

is arranged near the bottom of the chamber, which is supported upon legs, and the stove or fireplace is so arranged that fuel can be readily supplied to the grate, under which a pan is placed to receive the ashes. The products of combustion from the fire-box or grate, after circulating within the chamber, together with the impure air drawn from the apartment, pass away through a pipe or flue at the back of the chamber of the stove, such outlet pipe being provided with a

[1] *Vide Sanitary Record* for July, 1879.

damper. A coil of pipe is arranged within the interior of the chamber above the grate or fire-box—the lower end of this pipe passing out through the lower part of the case—and it is so arranged as to be in communication with the atmosphere exterior to the room or building to be heated. The upper end of this coil of pipe passes out at the top or upper part of the outer case of the stove, where it is in open communication with the interior of the room or building. The heat from the fire-grate portion of the stove is transmitted through the surface of the coil of pipe, and causes a current of air to circulate through the pipe from the exterior of the room into the interior, and this current of air has sufficient heat imparted to it without injury to its respiratory qualities by contact with the surface of the pipe, the thinness of the metal of which the pipe is formed causing it to transmit the heat rapidly without itself attaining a high temperature. A deposit of carbon is also formed upon the exterior of the pipe and upon the inner surface of the chamber, which assists in preventing the inlet pipe and the chamber from becoming overheated.

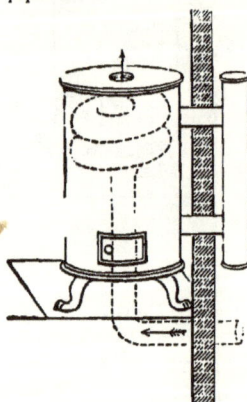

GEORGE'S GAS-BURNING CALORIGEN.

The success of the coal calorigen was so pronounced for use in rooms adapted for them, that in the year following, Mr. George gave his attention to the matter of gas stoves, and succeeded in producing what is called the gas calorigen.

In this contrivance, which is, like the coal stove, perfectly portable, the outer case is formed of thin sheet iron or copper, supported on legs; and a gas-burner of the ordinary ring form is fitted inside. The air for supporting combustion enters the chamber of the stove by the lower inlet pipe from the flue in communication with the chimney, and the products of combustion pass out of the chamber into the flue by the upper or

outlet pipe—an upward and downward current being thus established in the flue. There is also a door furnished with a panel
of glass or mica, which is closed when the gas is burning, so
as to exclude the passage of air through the opening of the
door. There is also a diaphragm, for distributing the air for
supporting combustion, and another diaphragm for causing the
heated products to circulate within the chamber of the stove.
It will thus be seen that the heat from the heated products of
combustion of gas burning within a chamber of metal is trans-

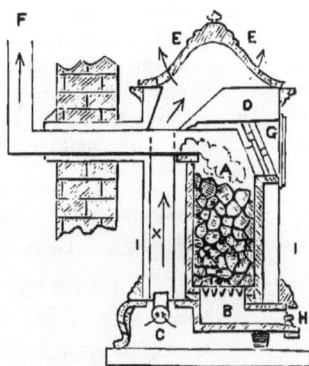
Fig. 1.

Fig. 2.

SLOW COMBUSTION CALORIGEN.
Fig. 1. Vertical Section. 2. Elevation. 3. Plan
through I. I.

Fig. 3.

A, fire chamber; B, ash-pit; C, fresh-air inlet; D, air
space; X, fresh-air chamber; E, warm fresh-air outlets; F, smoke outlet; G, feeding door; H, regulator.

mitted through and radiated from the sides of the chamber—
a current of air passing at the same time from the exterior to
the interior of the apartment—through a pipe of thin iron,
arranged in such a manner within the chamber of the stove as
to warm the air passing through it.

In 1879 a slow combustion calorigen was introduced, which
is well adapted for use in halls and vestibules where there is a

changing atmosphere. This stove, like others, is lighted in the usual manner upon a foundation of paper, small pieces of wood, and small coke, the ash door being removed to create a draught. When the coke is well alight, the stove is filled up with small coke and the ash door replaced. There is also a sliding or dropping valve introduced above the ash door so as to regulate the draught necessary for combustion.

As in the case of the coal and gas calorigens, these slow combustion calorigens are fitted up with a pipe, one end of which is in communication with the external atmosphere, and the other extremity terminating in the crescent-shaped warmed fresh-air chamber, the outlets of which are pierced in the surmounting of the stove. There is no escape of fumes when feeding, and at a cost of less than fourpence, the stove will burn, if properly alight and charged, without attention, for about twelve hours. It is therefore admirably adapted for keeping up a circulation of warm air in an inner hall or staircase during the night.

There are numerous other stoves constructed on much the same principle. In using these stoves care must be taken that the air to be warmed is drawn from a pure source, and any openings should be guarded from the entrance of vermin, etc., by double air-bricks or other means.

Water Supply.—In a village this will often be found a subject of no mean consideration. Should the hospital be near a small town which has a good water supply, and be so placed as to be easily connected with its mains, the difficulties will at once vanish, but in many villages the supply is *nil* or very bad.

The Rivers Pollution Commissioners classed the waters most fit for drinking and cooking purposes in the following order—(1) spring water, (2) deep well water, (3) upland surface water; while they condemned the use

of shallow well water, water from cultivated ground,
and rain water for the same purposes. It must always,
therefore, be the aim to get one of the three classes
first mentioned, treating, of course, each case on its
own merits. If the hospital can be placed in such a
position as to be able to utilise a good spring, such
a course must be adopted. Otherwise we must take
into consideration the geological features of the district, and the supply of water to places near at hand,
and then consider the propriety of sinking a well.
Dr. George Wilson, in his book on *Sanitary Depôts
in Villages*, says on the subject of water supply:—
"Specially suited for use in rural districts are the wells
known as Norton's Abyssinian tube wells. They consist of narrow iron tubes driven or screwed into the
ground in lengths, and with the lowest length pointed
and perforated at the end. The dangers arising from
the entrance of surface impurities are entirely obviated,
and they further possess the advantages of being
driven into any good water-bearing seam which may
be selected, of securing a sufficient yield in dry seasons,
and of entailing comparatively little outlay, either
for their first cost, or in sinking them. Wherever
pump wells are in use, these tube wells can be sunk.
In two days a well sixty feet deep can be sunk, which
in most cases will yield an abundant supply of pure
clear water within a few hours after completion." We
recommend these wells strongly; but, for the information of our readers, we must refer in addition to the
ordinary well. In no case, however, should recourse

be had to such a well until its local surroundings and physical conditions have been carefully examined, and the water from it and different wells in the neighbourhood has been examined and reported on by a competent analyst. The points to be attended to in the construction of a well are—(1) that it is not near any of the house drains, or any cesspool, or accumulation of filth; (2) that its mouth and sides are well protected from the entrance of surface water, liquid filth, etc.; (3) that proper materials are used in its construction. If this last point is left to a local builder, the water will probably be found to have been much hardened by the use of common mortar. In most cases the water will have to be pumped up, when it is usual to place a filter box at the top of the suction pipe leading to the well. The ordinary small filters here referred to, will however become so impure in the course of a week as to make the water fouler than it was when it entered them. If filtration is to be attempted, sand filters about 15 inches in depth, arranged in sets of three so that two can be drying while one is filtering the water, should be used. The water can, in an immense majority of cases, be pumped into a tank in the upper part of the building by means of either a windmill or a small gas engine, at a very small expense. If at any time there is a doubt about the purity of the water it can be rendered harmless by boiling. It will thus be seen how a supply of drinking and cooking water may in most cases be laid on for kitchen use. But what is to supply the bath-rooms and the closets (if there be any), should the hospital be built as we suggest in Chapter X.? It will entail a great amount of labour to pump a large supply to a cistern at the top of the house, and then to distribute it over the building by pipes. We may at any rate for the time leave out the question of closet supply and perhaps that of urinals, and consider how best to get a good stock of water for the bath-rooms and for other washing purposes, which shall be easy of access to the wards. Here the rain-water will be invaluable, not only because it can be readily utilised, but also because, though not fit for drinking, it is really the most valuable water for domestic purposes, and occupies in the report before mentioned the fore-most place amongst household waters. It can be easily stored in a galvanised iron or slate cistern under the roof over the bath-room, whence pipes may lead down to the bath and scullery. The cistern must be con-structed in such a manner that it can be easily cleaned at certain intervals, and the overflow pipe must end outside over a trapped grating. Means of ventilation must be provided; the entrance of leaves and other rubbish must be guarded against by fitting the pipes with perforated cups or roses, and it is also advisable that there should be a second smaller cistern into which the water may run through a char-coal filter before distribution. Taking the amount of rainfall in the year at 30 inches, and the roof space for each ward given in the model plan in Chapter X., the amount of rain-water will equal rather more than

10 gallons per head per day if the building is of a single story, but it will only equal about half that quantity if the building consist of two stories. By allowing the rain-water from the central administrative part (which will not be required in the kitchen) and from the corridors to flow into the ward cistern, we shall probably more than double the above figures, giving 20 gallons or 10 gallons per head per day respectively.[1] We have here been taking the most difficult case of supplying a large and somewhat scattered building. In a small purely cottage hospital the matter will be much simpler.

We must consider one other case. Hygienic improvement in many villages has not yet been even proposed ; and we may instance many large mining districts, where the mine workings extend in almost every direction, draining away most of the sources of water supply, so that there may be no springs, and it would be quite useless to think of sinking a well. Such

---

[1] As a matter of fact, it may be noted that during the winter there will almost always be an ample supply of rain-water for all these purposes. The great drawback to the utilisation of rain-water is that it generally fails in summer time ; but even then a good thunderstorm will often immediately fill the cistern to an extent equal to nearly a week's slow rainfall in winter. With a 35 in. rainfall 12 gals., and in the driest year 7·7 gals. per day

,,	40	,,	14·5 ,,	,,	,,	10·8 ,, ,,
,,	45	,,	16·4 ,,	,,	,,	13·1 ,, ,,

will be obtained, according to Mr. H. Sowerby Wallis, F.M.S. See his paper on the subject, published in the *Transactions of the Sanitary Institute of Great Britain* for 1879, page 210.

a case is by no means hypothetical, though it may seem so to many residents in towns. The inhabitants of these parts are chiefly dependent upon rain-water, and in a dry summer, when this fails, they are driven to make use of the water from the nearest brook, often only a running stream, contaminated in its course by sewage from each village or farm on its route, and by pumpings from the different mines. Or perhaps the villagers are able to utilise the water from the drainage of the neighbouring land—anything but a good supply—while the better class, who can afford to do so, will often cart their water for drinking purposes from a considerable distance. In such cases the use of water for closets is quite out of the question. The rain-water will have to be utilised as much as possible, always filtered before being used for drinking or cooking. The hospital must be built with as large a roof space as is compatible with the funds available; and a supply of "upland surface water" should be sought, care being taken that it does not come from cultivated ground. Perhaps the hospital may be fortunate enough to enlist the sympathies of a rich neighbour, who will allow his water-cart to bring it a supply of better water two or three times a week. This should then be carefully stored in some properly prepared receptacle. In such cases the water must be kept solely for drinking and cooking purposes, and it should always be filtered before use.

Should the water be very hard, Clark's softening process may be borne in mind as a useful auxiliary.

In arranging a system of hot water circulation the first and most important thing is to provide against the possibility of explosion. Almost every winter, with the advent of frost, disasters resulting from ill arranged systems occur, frequently with fatal results. The system which permits these accidents is arranged in this way. At the back of the kitchen range is a boiler which is connected by pipes with a hot water cistern at the top of the house, above which is a cold water supply cistern. The hot water therefore is drawn off from the highest point of the system, and it is possible to empty both cistern and pipes when by reason of the freezing of the water in the supply cistern there is no cold water to supply the place of that drawn off. The boiler becomes red hot, and if when in that condition a thaw occurs and cold water rushes in an explosion must take place. The system by which absolute safety can be assured is as follows. At the side of the kitchen range a hot water cylinder is placed, which is supplied with cold water from a cistern placed somewhere above the highest point to which hot water is required to be led. The cylinder is connected to the boiler by a flow and return pipe, and the supply of hot water to the various points is taken from the top of the cylinder. In case of frost the circulation of hot water ceases and the cylinder remains full or nearly full of water. Before the boiler could be emptied it would be necessary to evaporate the whole of the water in the cylinder, a process which would take a considerable time,—much longer indeed than any frost would be likely to last. An additional precaution of value is the provision of a safety valve properly weighted to the head of water, which is affixed to the cylinder. The cylinder should be provided with a manhole for cleansing and a valve for emptying and for economy of heat should be covered or "lagged" with a non-conducting composition.

Disposal of Excreta.—Under this head we must consider the best means of disposing of the excreta proper (fæces and urine), the slop and waste water, ashes, and the kitchen refuse. In most cases the rain-water, which otherwise would also have to be disposed of, will be utilised in another way, and may thus be left out of the present account.

Let us first consider the best means for getting rid of the fæces. The methods in common use in England are cesspits, water-closets, earth-closets, and ash-closets. The cesspit in all its varieties must be condemned unconditionally. It would be impossible to place one within easy access of the wards without creating a great nuisance, and it would require frequent disinfection and cleaning, so that in the end it would give as much trouble as an earth-closet, with none of the advantages of the latter. Ash-closets are not to be compared in efficiency with earth-closets, but they give the same amount of trouble. A combination of the two is, however, often admissible. The question, therefore, becomes narrowed down to a choice between earth-closets and water-closets. It is an old controversy which has gone on for

years, and is likely to continue. Indeed, it is
only practicable to point out here where the one
or the other, according to circumstances, may be
preferable. If there is a system of sewerage in force,
with which the hospital drains can be connected,
and if there is also a good supply of water, the water-
closet system must naturally be commended, and the
question becomes a simple one. The best method
for the disposal of sewage, if there is not a sufficient
water supply for water carriage, is by sub-surface
irrigation, the method of Mr. Bailey Denton and Mr.
Rogers Field, which is described further on. But in
most cases there will be no sewerage system in operation, and then all the arguments in favour of earth-
closets will press themselves strongly upon us. Thus,
there will be a large isolated building, with but limited
water supply, and requiring the purest possible air. The
earth-closet system can here be adopted at less cost
than water-closets, and with less liability to get out of
order ; and its adoption will also enable the rain-water
to be utilised for purposes of ablution, and will narrow
down the amount of drainage required. There is,
lastly, the further advantage that movable earth-
closets can be used in the wards for those patients
who are unable to leave their beds, so that the whole
system may be on one uniform plan.

Next, as to the working details of the earth-closet
system. The only drawback is the amount of labour
involved, and, in dealing with this difficulty, we must
enter a little into household arrangements. Con-

nected with the hospital there should be grounds
cheerfully laid out for the exercise of patients, and
perhaps a kitchen garden. In a country district this
will be one of the least difficulties. Land is there
cheap, and a good strip will usually be found connected with such a building as would be chosen for
conversion into a hospital. If a new hospital is being
erected, the site and a good piece of ground will probably have been given by the local magnate. To keep
the grounds in order will require the partial or entire
services of a man. It is needless to say that we urge
the employment of such a person about the place, and
we think there will be abundant employment for him
in innumerable ways. The want of a man about the
premises seems to have been much felt in several of
the larger hospitals, and the managers have been
obliged to make provision for the services of such
a person. At Boston and at Walsall, for example,
a man has been acknowledged to be indispensable,
and provision has been made for him by building a
lodge in the grounds. The man in such a case would
have charge of the garden and grounds, also of the
whole working of the earth-closet system, such as
the drying, renewing, and emptying of the earth : he
would be required to assist in carrying bad accident
cases into the wards, and in removing corpses to the
mortuary : he would help at operations, and act as
messenger or gate porter—in fact, such a person
would find abundant employment if care were taken
to choose for the post a man who is willing to make

himself generally useful.   He and his wife might perhaps occupy a cottage at the entrance to the grounds.
In that case they would act as gate porters, while a
wash-house might be built close to the cottage, where
the wife could manage the washing of the establishment.

The kind of earth used for closets is of importance. It should be of a loamy, porous nature,
as a sandy soil is of little use. It must also be well
sifted, otherwise it soon blocks up the orifices. In the
summer time it may be dried in the sun, and in winter
the kitchen fire may be utilised for the purpose, by
employing one of the cheap drying stoves made by
the Moule Closet Company. Wandering again from
the subject, we may point out how serviceable a gift a
small greenhouse would be from a generous donor.
Here the earth could be dried for the purposes of the
earth-closet, and the man would thus have the entire
management of this branch of work, without having
to intrude on kitchen territory, where he would be sure
to trample on the toes of the cook (or nurse). At an
extremely small cost, too, the wards could with this
adjunct be supplied with plants and flowers, and the
grounds be made bright and cheerful in the summer.
Hundreds of cuttings are thrown away yearly by rich
neighbours, who would be only too pleased to give
them for such a purpose ; and the cuttings once
obtained, a profusion of plants would, with a little
care, be forthcoming in a few years.

Whether the earth in the closet should be dried and

used again is a matter for consideration. Should it be important to make a good manure for purposes of sale, such a course could be adopted ; otherwise it is not very advisable, though no cases of injury from such a practice are recorded. We consider the use of disinfectants with the earth as needless, but, should they be deemed necessary, we would advise the use of chloralum rather than of carbolic acid or other strong smelling powder.

There are many good forms of earth-closet now in use ; the most common is that known as Moule's. The points to be attended to are,—that the apparatus be self-acting, that it be made to empty from the outside of the building, and that a pipe be led up from the receptacle to the outer air for purposes of ventilation. The quantity of earth required is about 1½ lb. per head daily. If a man is employed regularly about the hospital, the closets should be refilled and emptied early every morning, and the earth should then be stored in a shed for use or for removal. Mr. Moule, in a letter to the *Lancet*, October, 1867, cites the case of a school of 80 boys, in which the earth-closet system is in force, and the product though only removed once in three months never gives rise to any nuisance,—a farmer supplying and removing the earth. We simply give this as an instance of the little trouble occasioned by earth-closets, when it is difficult to procure labour, but cannot advise keeping the product near the wards in large receptacles for an equal length of time. Probably in most cases its removal once a week would be quite sufficient. The product could be sold to a farmer, or exchanged for fresh earth, or it might be used in the kitchen garden, should the hospital possess such a luxury. We have treated the subject of earth-closets simply as a system of convenience, without alluding to the monetary side of the question. It may be practicable to sell the product in rare cases, but the hospital authorities will generally have to be satisfied with an exchange of earth for manure, or its employment in the garden. *It is necessary to take care that the slops are not thrown down these closets.*

It is needless to enter into details as to the management of the water-closet system ; the rules are too well known. It will be sufficient to indicate the chief points concerning it. The water waste preventer must not be of less capacity than two gallons. The soil pipe must be carried above the roof and be left open at the top, and a ventilated manhole should be placed between the building and the sewer.

The principles of sound drainage are more generally known and recognised now than they were when the first edition of this book was published ; but the importance of absolute perfection in everything that concerns the drainage of a hospital is so great that no apology is needed for a recapitulation of the main rules to be observed.

Firstly, all drains or pipes conveying waste matter, whether above ground or below, must be absolutely water-tight in every part. Secondly, the whole system must be ventilated as freely as possible from end to end. Thirdly, the passage of air from the sewer or cesspool into the drains of the hospital must be prevented by the interposition of a water-seal or syphon trap at or near the point where the hospital drains join the sewer or cesspool. Fourthly, all waste pipes from baths, sinks, or lavatory basins, and all rain-water pipes, except where the latter are connected to a storage tank, must be made to discharge in the open air over open gulley traps.

Underground drains should be made of glazed stoneware pipes with socket joints. They should be laid in a bed of concrete to prevent injury to the joints by subsidence of the ground, and the joints should be made with neat cement, or with one of the patent bituminous joints run with Russian tallow and resin, and protected with cement. After being laid and jointed care must be taken to wipe out from each pipe any cement that may get squeezed out in process of jointing ; if this is not done the cement sets hard and forms a ridge inside the pipe against which solid matters lodge, and cause obstruction to the flow of sewage, frequently resulting in complete stoppage. When the drains are laid they are tested by plugging the lower end, and filling the pipes with water. The water is then allowed to stand for a sufficient period, to ensure that no leakage is taking place. The test is a severe one, but it is the only really accurate way of finding out if a drain is water-tight or not. The drains must be laid in straight lines from point to point ; and at every change of direction, or where junctions occur, manholes should be formed, through which the drains should run in open channels. A manhole is a rect-angular brick chamber or pit, varying in size according to its depth or the number of branch drains entering it ; its chief use is to afford ready access to the drains both for periodical inspection and for the removal of obstacles without having to break open the drain pipes. No drains should under any circumstances pass under any part of the building except where it may be necessary to drain the subsoil. In the latter case the subsoil drains should either be made to discharge into an open watercourse or ditch, or, if that is imprac-ticable, they must be disconnected by means of an open water-seal trap from the sewage drains.

The vertical pipes may either be of heavy cast iron or of lead. For soil pipes lead is preferable, but it must be of not less weight than eight pounds to the superficial foot, and must be what is known as " drawn " pipe, that is, without a vertical joint or seam. For both waste pipes and sink pipes where hot water is liable to be discharged, heavy socketed iron piping is more suitable, as the expansion and contraction are much less than with the lead pipe. Soil pipes should be connected at their feet directly to the drain without any trap, and should be carried up well above all windows full bore as ventilating shafts. Waste pipes should also be carried up as ventilating shafts, but must, as before explained, discharge over trapped gulleys.

Such, in brief outline, are the main principles of

sound drainage, the two chief points to be borne in mind being the necessity for simply planned direct lines of absolutely water-tight pipes, and the provision of free ventilation to all parts of the system.

The annexed diagram shows the mode of constructing a disconnecting manhole. In size it will vary with the depth of the drain from the surface,

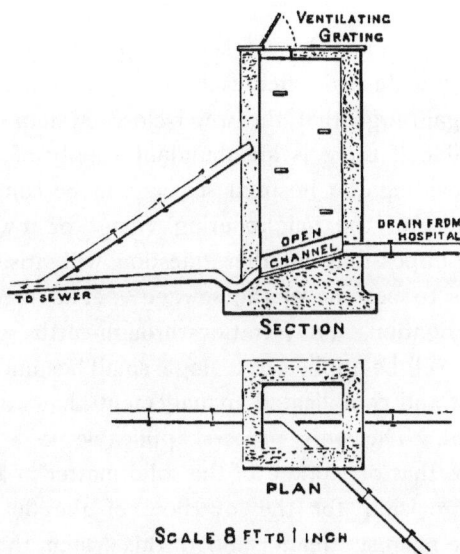

SCALE 8 FT TO 1 INCH

a very deep manhole requiring to be considerably larger than a shallow one in order to allow sufficient room for working the drain rods should a stoppage occur.

Where water-closets are used the apparatus should be of the simplest possible kind. The form known as the "Pedestal Hygienic" is one of the best. It is made in one piece of white porcelain, and with a three-gallon flush is as nearly self-cleansing as any closet can be. To each closet should be provided a flushing cistern, which will discharge automatically three gallons of water on the handle being pulled. The water supply pipes here, as in all other parts of the hospital, should be fixed clear of the face of the wall, and in no case concealed in the plaster or behind casings.

The pan must be of such a pattern as to be thoroughly flushed after each occasion of use. We must again urge that the water-closet system is only admissible if there is an abundant supply of water, and providing the hospital sewers can be connected with those of the neighbouring village or town. If this is impossible, another question will arise, *viz.*, what is to become of the sewage? The precipitation, irrigation, and filtration-through-earth systems cannot well be applied to a single small hospital, from the cost and complicated management that would be required. The only process applicable to a single house is that of storage of the solid matter in a tank, with provision for the overflow of the fluid part into the nearest stream ; and to this system there are many obvious disadvantages. Nor is it legal or right to pollute the neighbouring brooks, by discharging the sewage into them at once, which would be the most simple plan of cutting the Gordian knot. We may mention, however, two other modes, which have been recommended as applicable on a small scale.

The first is Mr. Taylor's plan of collecting the fæces separately from the urine,—the fæces, mixed with earth and disinfected, to form guano, and the urine to be discharged, as he considers it of little use for manure. Similar plans, having much the same object, have been advised by others, but they differ much in detail; for while some consider the urine to be of little value as a fertilising agent, others consider that it is the more important of the two. Until some uniform plan, therefore, has been agreed upon, it is impossible to recommend this system. In the *Lancet* for June, 1875, is described a syphon tank, designed by Mr. Bailey Denton and Mr. Rogers Field, specially for the use of villages and isolated buildings. This was strongly recommended also by the late Mr. Netten Radcliffe and Dr. George Wilson, and would seem to answer every purpose, and to be a very complete contrivance. It requires, however, land for the reception of the sewage, in the form either of a kitchen garden, or of some farm close at hand, with a farmer who could make the sewage a source of profit, and would undertake the management of the arrangements.[1]

Before leaving the subject of closets, one obvious consideration will force itself upon us. Are enteric fever cases to be admitted into the hospital, and if

[1] Dr. Wilson says, when these tanks are used, "½ an acre to ¾ of an acre of ground properly drained and laid out would be quite sufficient to purify the slops and refuse water of a village of 800 to 1000 inhabitants, provided the subsoil is porous."

so, what precautions are to be taken with regard to their stools? Without entering into a long and doubtful controversy, let us state at once that we regard enteric fever as a specific disease, the germs of which are only disseminated by means of enteric fever stools; and the majority of the profession will probably assent to this. Moreover, enteric fever cases are admitted into general wards, and, in fever hospitals, other acute cases, which find their way thither by mistake, are treated in the enteric fever wards. In fact, enteric fever is always treated as a non-contagious disease, where the stools are properly disinfected, and it is very rarely communicated from one patient to another or to the nurses. But, on the other hand, no cases require more care as regards diet and nursing than cases of this disease. In many villages also, where the sanitary arrangements are very defective, we find enteric fever of a bad type constantly breaking out ; and as the country medical practitioners often teach that it is a most infectious disease, all the friends and neighbours desert the patient in his or her sorest need, and no one can be prevailed upon to come and nurse the sufferer. This is no hypothetical case. We therefore urge that, at any rate, enteric fever cases should be received into the cottage hospital, but that their stools should be kept quite separate from those of the others, that the motions should be passed into a vessel containing some strong disinfectant, that they should be immediately covered with another layer of the powder, and that they should be removed daily or twice a day, and buried deeply in the ground at a distance from any well, drain, etc. With these precautions no harm will be at all likely to ensue.

Of course, where proper provision has been made by the local sanitary authority for the isolation of infectious diseases, the need for the treatment of ordinary cases of enteric fever in the cottage hospital ought not to arise. But unfortunately in too large a proportion of our rural districts the local authorities have not made any proper isolation hospital provision, and then the fact that enteric cases can be safely treated in the cottage hospital, if the simple precautions we have indicated are observed, will prove an inestimable boon to many a poor cottager as well as to the whole community.

Urinals and Disposal of Slops.—Many good forms of urinals are now in use, so that there need be no difficulty in choosing a suitable pattern. They are usually made of earthenware or enamelled iron. Those formed of large slabs of slate, which give off a most offensive smell from the large area of surface exposed, must certainly be condemned. Care should always be taken to choose a pattern with a proper receiver. In case of a good supply of water, it may be used for flushing the urinals, and the best plan is to have it so arranged that a strong stream may be turned on twice or thrice daily, if the nurse can be trusted to do so efficiently. Otherwise the urinals must be made self-acting, though this would entail a larger use of water with less efficiency. If water cannot be spared for the purpose, the dry-earth system may be adopted ; and the apparatus may either be made self-acting (several forms of this kind are now made), or the earth may be supplied by hand daily. But in many cases there may be a prejudice against applying the dry-earth system to the urinals. Such a prejudice certainly does exist, probably because it is not a common thing to treat the urinary excretion in that manner. Under these circumstances, it will be best to collect the urine and slop-water together, utilising them in some manner if possible ; and should there be a garden attached, improvements in this direction have lately been so great that it can be done easily, cheaply, and effectively. An excellent plan has been invented by Mr. Bailey Denton and Mr. Rogers Field, to which allusion has already been made. The water, urine, etc., are here led off into a tank, from which they are allowed to escape automatically, either over land already prepared to receive them, or into sub-irrigation drains, laid below the surface of the garden, while any deposit is removed from the bottom of the tank once a month. The following description of this system is given by Mr. Netten Radcliffe, in the Second Report of the Medical Officer of the Local Government Board for 1874 (pages 232-235) :—

The apparatus shown in section in the figure (page 197) consists of a cylindrical water-tight iron tank (A), having a trapped inlet (B), which also forms a movable cover to give access to the inside of the tank, and a socket (C) for a ventilating pipe. The outlet consists of a syphon (DEF), so arranged that no discharge takes place till the tank is completely filled with sewage, when the syphon is brought into action and the contents are immediately discharged. The outer end (F) of the syphon dips into a discharging trough (G), attached to the flange of the syphon by a movable button (H), so as to be turned round in the right direction, to connect the tank with the line of outlet pipes (I). This trough has a barrier (J) across it, with a notch so contrived as to assist small quantities of liquid in bringing the syphon into action, instead of merely dribbling over the syphon without charging it, as they other-

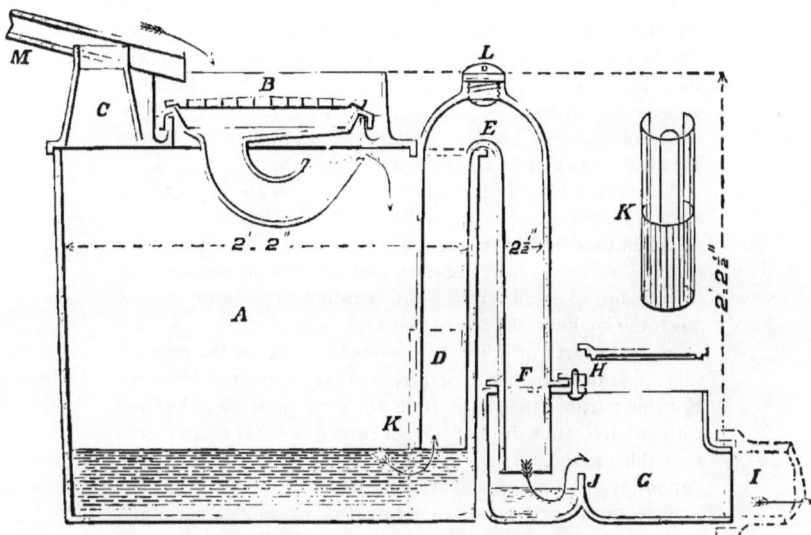

FIELD'S SELF-ACTING FLUSH TANK.

wise would do. The cover of this trough can be removed to give access for cleaning.

There is also a brass-wire strainer (K), which is clipped on to the inner end (D) of the syphon, so as to be taken off at will ; and a screwed brass plug (L) is fitted to the bend (E) of the syphon in case it should at any time be necessary to examine or clear it. The pipe (M) represents a waste-water pipe (usually from a sink) through which the supply of sewage is conveyed to the tank.

When used for *flushing drains*, all that is required is to fix

the tank outside the house or building and in some convenient
position between the supply and the drain to be flushed, and to
connect the supply with the inlet and the drain with the outlet
of the tank.   There is no house in which there is not sufficient
waste water for flushing by means of this apparatus.   The sink
or scullery slops are generally available as a supply, and the
tank is especially adapted for them, as it forms the most perfect
kind of trap, breaking the connection between the drains and
the house and intercepting the fat.   Where the drains have
only slight fall, advantage can be taken of the height of the
sink by placing the top of the tank above the ground.   The
drippings from a water-tap, or the rain-water from a roof, may
also be used as a supply.   A very small accession of water will
start the syphon when the tank is once full, but, should it
occasionally remain full for any time in consequence of insufficient supply, a jug of water thrown on the grating of the
inlet will immediately set the syphon in action.

The tank holds about 40 gallons and delivers about 33 gallons
at each discharge.

When used for the *disposal of house slops, where no regular
system of sewerage exists*, the flush tank enables all house refuse
to be removed inoffensively—the bedroom slops being thrown
down the basin at the top of the tank outside the house—and
thus where earth or other dry closets are used for the excreta,
this apparatus supplies a complete sanitary system of drainage.
For this purpose the outlet from the tank must be connected
with sub-irrigation drains laid in a garden or other small plot of
available ground.   These drains may consist of common 2-inch
agricultural drain pipes, laid some 10 or 12 inches below the
surface, on a continuous bed formed by dividing larger pipes,

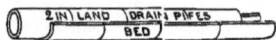

longitudinally, into two equal
parts, as shown in the adjoining figure.   This bed is
not disturbed when the pipes
are taken up to be cleaned,
are taken up to be cleaned,

METHOD OF LAYING THE SUB-
IRRIGATION DRAINS.

and thus ensures their always being relaid in true position.
The sewage flows out of the joints into the soil, and feeds the
vegetation, and the concentration of the flow effected by the

sudden discharge of the tank forces the liquid rapidly along the pipes and prevents their being choked. The sewage can thus be distributed over a sufficient area of ground to give full opportunity for the soil to purify it on the principle of "intermittent downward filtration."

By this method of disposing of slops the difficulty of sub-irrigation by gravitation from the ordinary flow is overcome.

It will be observed in the figure on page 197 that the sink drain opens outside the tank above a trapped inlet; and that the long limb of the syphon terminates in a small trough. This trough is so arranged that any trickle of slops along the syphon quickly closes the outlet; and continuation of the trickle after this closure, from air contained in the syphon being carried along with the liquid, exhausts the syphon sufficiently to cause it to be brought into action by the preponderating atmospheric pressure upon the surface of the liquid in the tank. By this ingenious arrangement a much less quantity of liquid thrown down the sink will start this syphon in action, than in the tank first designed, and abortive flow from the tank is obviated.

The flush tank is made of different sizes, varying in capacity from 16 to 30 gallons, exclusive of space for deposit. When in continuous use the interior should be cleansed and the deposit (a useful manure) removed every month.

The foregoing invention for separate houses has been further developed so as to meet the general wants of a community, by Mr. Bailey Denton, C.E., in conjunction with Mr. Field. The same difficulties have been experienced in dealing with and utilising readily, and economically, and inoffensively the liquid house refuse of villages and towns, as with that of separate houses, and from the same causes, the ordinary insignificance and occasional irregularities of flow. To meet these difficulties Messrs. Bailey Denton and Field have designed a tank termed by them "The Automatic Sewage Meter," which is constructed on the same principles as Mr. Field's self-acting flush tank. This meter provides for the accumulation of the liquid refuse and for its automatic discharge, at definite intervals, in quantities admitting of distribution over land by gravitation, for purposes of irrigation. A meter of this kind has been in use some years in the hamlet of Eastwick, near Leatherhead in Surrey; and its operation, particularly as part of the sanitary arrangements of the village, may there be studied very usefully.

Eastwick at the time the system was adopted was a hamlet of thirteen houses, including the mansion of the proprietor and the farm homestead; and it had a population of about 145. In devising a system of excrement and slop disposal for the place, any general plan for water sewerage had to be set aside, the water supply derived from wells being variable in quantity and at no time too abundant for ordinary domestic use, irrespective of water-closets. The common privy was retained for the cottages, but the privy pit was converted into a water-tight receptacle beneath the floor of the closet, and the cottagers were instructed to throw into it above the excrement the refuse ashes, and to remove the contents of the pit monthly for use in their gardens. Four water-closets existed, and five earth-closets, for the use of the mansion and its precincts; and one water-closet and three earth-closets for the use of the farm homestead. To provide for the liquid house refuse of the hamlet and for the drainage of the farm buildings, a scheme of sewerage was carried out by Mr. Bailey Denton, which had an outlet in a meter tank. This tank was in two compartments, to admit of cleansing without entire disuse. It had a capacity of 500 gallons, and it filled and discharged in ordinary dry weather three times in two days. The several discharges were directed successively on different portions of a plot of ground prepared for the purpose, and which, measuring 3 roods 3 perches, served ordinarily for the effective and profitable utilisation of the whole liquid refuse of the several cottages, the mansion, and the farmstead. The drainage of the latter included the flow from cattle sheds and stables, in which from 15 to 20 animals were always present, and about 30 head of horned cattle and 30 horses at intervals. The drainage of a large piggery also passed to the tanks.

Mr. Bailey Denton has furnished the following statement of the cost of the works above described, including the " meter " and the preparation of the land ; and he has remarked upon this statement that "the yearly return, after deducting the cost of attendance upon the sewaged land and regulator, cannot be less than £17 per annum, so that a return of about five per cent. on the outlay was gained, while there was every prospect of increasing that return, as the quantity of sewage dealt with became greater, and its treatment became better understood."

EASTWICK SEWERAGE.

			£	s.	d.
To payment for labour,	.	. . .	179	4	0
,,	,, pipes, .	. . . .	103	7	2
,,	,, stone, lime, cement and sand,		12	14	0
,,	,, iron and lead work,	. .	20	5	1
,,	,, carriage of materials, .	.	1	9	1
Travelling and incidental expenses,		. .	3	12	0
			£320	11	4

In regard to abatement of slop nuisance, and it may be added also largely of farm nuisance, among a rural community, the arrangements at Eastwick were the most complete and satisfactory that Mr. Radcliffe had seen. Notwithstanding the contiguity of the irrigated land to the mansion, no nuisance was experienced from it, whereas previous to the above arrangements, when the slops of the mansion and cottages found their way into the neighbouring ditches and decomposed there, considerable nuisance had existed.

As the treatment of slop water and urine may so far be managed together, we will finish this branch of the subject by mentioning two other plans in use at different places, as there may be cases in which one or the other may be found more feasible than the foregoing. Thus, in one convalescent hospital, where the dry-earth system is in force, the refuse sink water is drained into a carefully-made cesspool, easily accessible for a cart. The contents of this cesspool are pumped out and removed daily, being used by a market gardener as manure. Another plan is to run the sewage into proper earth tanks, when all smell is avoided, and the soil is enriched in a slight degree. Dr. George Wilson writes on this point, and recommends the following simple plan for adoption, having himself tried it with success : —

" I recommend that the sewage, if it be possible, should be purified by irrigation or sub-irrigation, and, failing these, that it should be filtered through a sand and gravel filter of sufficient size, and on the intermittent downward filtration principle. In some localities, where the soil is porous, the outfall drain may be carried alongside a field ditch of lower depth, and the soil between will act as a ready filter. Indeed, in purely agricultural districts the various expedients which might easily be adopted are so accessible, so to speak, that in the great majority of instances there not only need be no difficulty in treating them efficiently, but, if properly utilised, they (the village slops) will pay a fair return to any farmer who is public-spirited enough to take them."

As regards the *kitchen refuse* (potato-peelings, cabbage-leaves, etc., etc.), many people in the country will be only too glad to buy and remove it for their pigs. One condition should be made with them—that they remove it daily at a stated time. It may be shot from the kitchen into a properly made receptacle, so ventilated as to prevent the return of bad air into the kitchen, and having an opening outside by means of which it can be emptied from the exterior.

A properly contrived bin for ashes will also be needed. These will be found very useful for repairing the walks in the kitchen garden ; or, mixed with earth and sifted, they may be used for the earth-closets. The cinder-sifters invented by Mr. Fox of Cockermouth and Dr. Bond of Gloucester are recommended for adoption. The drain-pipes, waste-pipes, etc., will want careful attention, especially if it is determined to make use of an old building. Each must be carefully inspected to see that it is properly trapped ; that it runs outside the building, in such a position that it can be easily inspected ; that the joints are perfect ; that it is disconnected from the sewers by ending over a trapped grating ; and that it is properly ventilated by a pipe carried above the roof. The overflow pipe from the cistern will require special attention.

We may almost consider after all, that a country cottage or farmhouse, with a small piece of kitchen garden, ought to be more easily and economically managed, as regards the utilisation of its sanitary products, than a building having the complex machinery required in dealing with large towns. Thus we first utilise the rain-water ; next, the fæces and urine, by mixing them with earth, and employing them as manure in the garden, or by selling the produce to the neighbouring farmers ; the kitchen refuse again makes a valuable food for pigs ; and the ashes are used up in repairing the garden walks or in the earth-closets. What then remains? Nothing but the bath and slop water, which can easily and scientifically be got rid of by the means described, if it is allowed to flow with the other liquid sewage into Denton and Field's tank.

We will conclude with some remarks on the subject of slops quoted from Dr. Wilson's little book, which ought to find its way into the hands of every country medical practitioner :—

"Coming now to the question of the disposal of liquid refuse, I would premise at the outset that whether slops are or are not sewage in the legal sense of the term, they often give rise to serious nuisance, and by polluting wells or other sources of water supply, they are frequently the cause of serious illness. They must, therefore, be disposed of in some satisfactory way. It has been assumed that, if you prevent any admixture of excremental filth, slop water is comparatively harmless, and may be allowed to flow into a canal or water-course without detriment. But there is one instance in my own district (Mid War-wickshire) in which, apart altogether from nuisance, a canal company has obtained an injunction against a rural sanitary authority, for silting up a part of the canal with the deposit of slop water from a village which discharges into it ; and there are several instances in which I have considered it imperative that village slops should be purified before they enter streams which are used as sources of water supply below the villages."

We quote the above paragraph in proof of our assertion, that the old notion with regard to the harmless character of slops, and the trifling importance, from a sanitary point of view, of any special care being taken as to their disposal, is disproved by experience. And since of all people the cottage hospital managers ought to be as perfect as possible, so far as the sanitary arrangements of their institutions are concerned, they ought to be very careful not to fail in this respect by a too easy disregard of trifles, wrongly so called. Should the cottage hospital be isolated, and means of drainage of a complete character be consequently deficient, the slops should be kept in one of Field's syphon flush tanks, from which they can easily be transferred for use in the garden. Dr. Bond's precipitating slop-tub can also be recommended with confidence, as by its means the slops are sure to be properly filtered.

CHAPTER IX.

A MORE DETAILED ACCOUNT OF CERTAIN COTTAGE
HOSPITALS, WITH PLANS AND ELEVATIONS.

Description of the Cottage Hospitals at Cranleigh, Bourton-
on-the-Water, Harrow, Petersfield, Harrogate, Lynton,
Ditchingham, Leek, Walker, Ashburton and Buckfast-
leigh, Redruth, Petworth, Wirksworth, Reigate, Boston,
Sherborne. Braintree and Bocking—Chronic Hospitals—
Rules at Saffron Walden.

THE cottage hospital movement has undergone such
a vast development since the issue of Dr. Swete's
work that it is now impossible to treat each hospital
in detail. We have therefore picked out a few from
the general list which have certain peculiar points
about them worthy of note. We treat them in detail
in certain particulars, which do not appear so fully
in the tables and elsewhere.

Cranleigh Village Hospital.—This was the first
cottage hospital, and was commenced in the year 1859
by the late Mr. Napper, who has been appropriately
called the father of the movement. It still remains,
save for a few slight alterations, in the same
state in which it was first started,—not, however,
from any want of funds. On the contrary, Mr.

(206)

Napper was offered an entirely new hospital; but by his wish the original hospital stands as a memorial of the first of its kind,—as an instance of what can be done with very slight means. With these sentiments every one must agree, and it would be a sad pity to do away with what is now so distinguished a landmark in the history of this movement. The building is a very old cottage of the ordinary Surrey type, given by the rector free of rent, and adapted by alterations and fittings for its present use at a cost of about £50. The simple but effective manner in which the alterations have been carried out is worthy of all praise, and displays much ingenuity. The alterations ought to be seen by every one interested in the erection of a cottage hospital, as they show by what simple means almost every obstacle can be surmounted. We must particularly mention the ingenuity displayed in the construction of a bathroom, with a supply of hot and cold water. The nursing is placed in charge of a trained nurse under the supervision of a ladies' committee. Much has been said (and the rule has been extensively copied into those of other cottage hospitals) about the visiting of patients at their own homes by the nurse. Now the late Mr. Napper stated to the author that this rule had never been carried out here, for the simple reason that the nurse never had time for such extra duties. In making such a rule, Mr. Napper explained that it was his idea, that at some future time a second nurse might be found requisite; and should such be the case, the latter might in her spare time help to

COTTAGE HOSPITAL CRANLEIGH SURREY.

SCALE OF   10   5   0   10   20   30   40   50   FEET

ROOF

BED   BED

OPERATION ROOM

BED

ROOF   WARD   16'0"   WARD   NURSES ROOM

ROOF

STORE RM

11'0"   12'0"

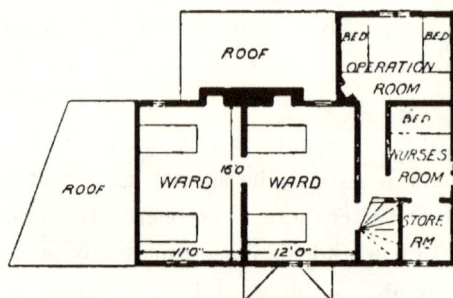

FIRST FLOOR PLAN

THE BUILDING CONTAINS
    GROUND FLOOR
    FIRST   "

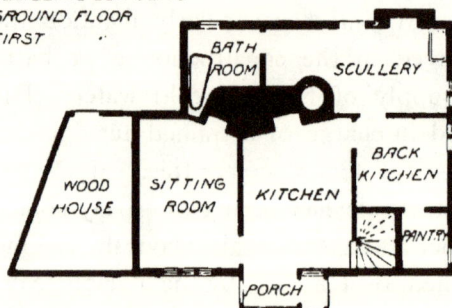

BATH ROOM

SCULLERY

BACK KITCHEN

WOOD HOUSE   SITTING ROOM   KITCHEN

PANTRY

PORCH

GROUND FLOOR PLAN

nurse the sick in their own homes. This system has
been found to work well in practice. But if with a

yearly average of only about 30 patients, one nurse
is fully employed, surely with over 60 patients, a
hospital must require the sole and continual services
of one nurse.

Bourton-on-the-Water.—This, the third cottage
hospital opened, was established in 1861, chiefly by
the exertions of Mr. John Moore, M.R.C.S. From an
interesting little pamphlet by the late Dr. Coles,[1]
we learn that the house in which the hospital was
first located was a substantial, but not a modern
building, of three stories, and hence necessarily
the rooms were low. It was doubtless erected on
the plan, adopted generally in the village about a
century ago, of utilising all available space with the
least possible expenditure of money. The lower
rooms on the basement floor were a little below the
surface level on the outside, so that you descended a
step to enter the house. It had been added to, and
adapted, as circumstances called for, to meet particular requirements, by special funds raised and set
apart for this purpose. It was the home of a pensioned
soldier and his wife, the latter acting as nurse. It was
situated on the outskirts of Bourton, but was easily
accessible, had a southern aspect, and from its upper
windows commanded pleasant views over the surrounding country. There was a garden, in which convalescents took exercise, whilst its products were found

---

[1] *A Rural Village Hospital,* by W. B. Coles. London : Taylor
and Co. 1877.

14

useful in the hospital dietary. The want of a good
carriage approach was greatly felt, insomuch that a
vehicle conveying an injured person could not draw
up close to its principal entrance-door.

This building was in use from 1861 up to the
middle of the year 1879, and was of incalculable
service, 665 in-patients having been treated under
its roof during the eighteen years of its existence,
and upwards of 4000 out-patients having resorted
to it.

Having regard, however, to its inconvenient situation and faulty arrangement, and to its low and confined rooms and yearly increasing dilapidations, it was
felt necessary in 1879 to find some more suitable premises, instead of expending any further money in trying
to improve the existing building. A site was generously
given by a resident, other contributions came in, and
the result of the agitation was so successful, that in the
autumn of 1879 an entirely new and compact hospital, erected of brick and tiles, at a cost of £887,
was opened with appropriate ceremonies, and was
duly placed at the service of the village. The property has been fenced in with a light iron railing,
and trees and shrubs have been planted.

In planning the hospital, the first consideration,
after its probable cost, was to provide sufficient cubical
space for its inmates ; the next to secure increased
facilities for efficient treatment and nursing. In the
old hospital, the want of adequate cubic space, and of
other proper sanitary arrangements, had, by retarding

convalescence, necessitated a protracted stay and consequently increased expenditure. During the past few years a complete system of drainage has been established, an efficient system of heating the hospital with hot water has been fitted, and a trained matron has been engaged.

Harrow Cottage Hospital.—From the bye-laws of this hospital being most complete and elaborate, we are induced to quote them much in detail, as they may be of great service in drawing up a set of rules and bye-laws for a newly started cottage hospital.

Subscribers and Donors.

All subscribers are entitled to recommend patients, who shall be admitted on a doctor's certificate on payment of the usual fees. No subscriber shall have the power of recommending patients in any year beginning 1st January until his subscription for that year shall have been paid.

Ladies' Committee.

[We reproduce these in full, on p. 77.]

In-Patients.

Letter of recommendation required (except in cases of accident or emergency), which must be countersigned by a medical man, that the case is likely to be benefited by such admission.

In cases of accident the nurse may give provisional admission, but must report it at once to the medical officer.

Payments—labourers, 9d.; under 12 years of age, 6d.; small shopkeepers or mechanics, 1s.; domestic servants in service, 2s.; out of service, 1s.; for each night they remain in the hospital. Each patient must bring a paper signed by some one guaranteeing this payment.

Rules as to the proper conduct of the patients, and as to the assistance required to be given to the nurse by such as are able.

Each patient will be held responsible for the washing of his own clothes.

Visits from patients' friends allowed on Wednesdays and Saturdays; from district visitors on Tuesdays and Thursdays.

Out-Patients.

If these attend at the hospital, they pay 1s. 6d. for four weeks. If attended at their own homes they pay 2s. 6d. for two weeks, subject to deduction if they are only seen once at home, and afterwards ordered to attend at the hospital. Pauper patients are not to be treated as out-patients.

Medical Officers

will visit in-patients daily. They will recommend to the board the medicines it is necessary to purchase. They will keep a journal accessible at all times to the board of managers, in which will be entered the name, occupation, residence, date of admission and discharge of each patient, names of persons recommending, as also the result of their treatment.

Nurse.

Rules as to (1) proper nursing of patients; (2) proper washing of clothes, disinfection and ventilation; (3) administration of diets; (4) superintendence of cooking, etc.; (5) administration of medicines; (6) reading of prayers; (7) care of books, and visits of ministers of all denominations when desired by the patient; (8) proper weighing and measuring of all provisions brought to the hospital, and superintendence of accounts.

These rules are too long to quote in their entirety, but it will be seen from the above headings and entries, that the system of management in force is most exhaustive as to detail, and we recommend any one interested in this movement to write for a copy of the rules in their complete form.

Petersfield Cottage Hospital.—The following are extracts from a letter from Mr. T. Moore, the consulting surgeon and late honorary secretary :[1] "We have four wards, two with two beds, two with one bed on the ground floor, with kitchen, scullery, matron's, and operating room. Upstairs, a convalescent room and a nurse's room. The operating room was originally upstairs, but so much difficulty was experienced in getting patients down to the wards after operations, that one has been built on the ground floor. It is, I believe, a great point to have all the wards, kitchen, etc., on one floor, as much labour is saved thereby. A skylight in the operating room is a *sine quâ non.* The walls are lined with Parian cement, which is easily cleaned. We have Galton's ventilating stoves ; they are ugly, but good, I think. Our windows open on hinges— a great mistake, for they let in the wind and wet. If I built again, I would insist on sash windows and plate glass ; we have the present windows because our committee thought they looked more 'cottagy,' and crown glass because it is cheap. We talk of altering them, when we can get money to do it.

[1] These particulars are as they appeared in the second edition, no reply having been received to our inquiries.

We have earth-closets, which act very well when they are well supplied with dry earth, but it is difficult to get that done. We have found it almost impossible to do without a mortuary, and have just erected a small detached building for that purpose. We have a head nurse or matron at £20 a year and £4 for beer, and a girl at £11 and £2 for beer. All the medical men in the town are medical officers, but none of those in the villages round ; the medicines are dispensed by a druggist in town at so much per bottle.

" The wards for men and women are at opposite ends of the building, and are separated from one another by the kitchen and other offices, and shut off by doors. Upstairs is a convalescent room and a nurse's bedroom.

" We make it a rule that the matron shall take her meals with the convalescent patients, and help all the others ; and this we find certainly saves waste. Whilst insisting on economical management, we never spare expenses for anything considered necessary in the way of diet, surgical appliances, etc., for we believe that liberality in this respect is in the long run the best policy."

The above is a most interesting experience in many little practical points.

Mr. Moore is a surgeon of no mean ability. He is always on the outlook for improvements, and every new system or appliance or remedy is carefully tested at Petersfield. A visit to the hospital has more than confirmed the favourable opinion we formed of its completeness. Its management is a model in many ways, the regularity of the attendance of the house visitors is beyond praise, and the systematic and accurate book-keeping is worthy of imitation. The hospital case-book, as designed by Mr. Moore, is the most complete we have seen, and every cottage hospital should possess a similar one. A sample leaf will be found at the end of this book (p. 358).

Harrogate Cottage Hospital.—First established in 1870 in two cottages, which were rented at £26 per annum, and in 1873 removed to two small houses which were purchased by the committee for £550, and adapted to the requirements of a hospital at a further cost of about £300, both sums being raised by subscription. In 1883 a new hospital, having 25 beds, was erected on a site adjoining the old premises. It 1895 it was decided to still further enlarge the hospital, giving further accommodation for 17 adults and 8 children. This enlargement is now being carried out at a cost of £2000.

There are three honorary medical officers, a matron, two nurses, and a domestic staff of three servants and a porter.

Lynton District Cottage Hospital.—The site of the hospital was given for 99 years by the late Mr. James Lean; the building was commenced in April, 1873, and opened in May, 1874. Its cost, including the furnishing, was about £680, besides which there were gifts of stone and some timber. Much carting was required, owing to the nature of the ground, and it was done gratuitously by the farmers of the district. This was reduced to a money value; and taking a cart with one horse at 5s., with two horses at 8s. per day, the total amount was found to come to £77 10s. 6d., which sum is included in the above £680. There is accommodation for five patients. The institution has been extremely successful and popular, owing perhaps in part to its distance from the county hospital, but much to the good management. We commend the two following rules :—

> Admission of patients to the hospital shall be granted by the medical officer, such admissions, and the retention of the patients in the hospital, being subject to the approval of the committee.
>
> Patients shall be received without forms of recommendation, and shall contribute towards their maintenance a weekly sum, the amount of which shall be fixed by the medical officer, subject to the approval of the committee.

As a rule, pauper cases pay nothing; servants, labourers, and artisans from 2s. to 5s. a week, according to income, number in family, and whether in a sick club or not.

Ditchingham—All Hallows Country Hospital.[1]— This hospital was opened in July, 1873, having cost about £3000. It is a large, rather rambling, one-storied building on the pavilion principle. Besides the ordinary wards, it contains special

---

[1] These particulars are as they appeared in the second edition, no reply having been received to our inquiries.

wards, an operating room, bath room, chapel, etc.
There are two surgeons, and a consulting surgeon
residing in Norwich.   Patients of the poorest class
pay 5s. weekly, domestic servants, 7s. 6d., persons in
better circumstances, 10s., as also incurable or permanent patients.   The nursing is under the care of
the All Saints Sisters of Mercy, by whom the hospital
was built, and the head sister is called "sister in
charge."

Leek Memorial Cottage Hospital.—This hospital
was built and furnished by Mrs. Alsop in 1870, in
memory of her late husband, and was managed and
maintained entirely by her until May, 1874, when a
meeting was called to consider a plan for its future
management and support.   This first meeting met
with but a faint response, and the affair seemed inclined to collapse ; but the trustees in another con-
ference with the medical men of the town, determined
to open it as a public charitable institution, and to
look to the public for support.   Mrs. Alsop agreed
to this, and gave up the hospital to a committee for
three years, keeping the building in repair for that
time, and promising to give £100 for the first year,
£50 for the second year, and £25 for the third year,
and if at the expiration of that time it was a success,
to hand the building over to the committee altogether.
The hospital thus constituted was opened again on
May 4, 1874, as a public institution under the provisions of a trust deed drawn up under Mrs. Alsop's
directions.   A new trust deed was prepared in 1889,

and under its provisions the hospital was governed until the end of 1894. Considerable differences of opinion having existed as to the principles of government, a new scheme was put in force in February, 1895. Letters of recommendation are required, and patients pay not less than 2s. 6d. weekly, except those provided with a church, chapel, or work-people's recommendation, who are admitted free.

The hospital is pleasantly situated on the borders of the town, and its original site of 830 square yards has been recently enlarged by the gift and purchase of about 1500 square yards of land. On the ground floor are nurses' rooms, kitchen, scullery, pantry, male ward (four beds), single-bedded male ward, and male bath rooms and closets. On first floor, female ward (three beds and two cots), single-bedded female ward, female bath rooms and closets, operation room, nurse's bedroom, and above this the servants' room. The outhouses include mortuary, dust bin, coal hole, etc. The wards are spacious and airy, warmed by fireplaces, etc., and also in the winter by hot-air flues, the fire of which is lighted from the outside ; the inlets are arranged along the floor, and the outlets near the ceiling into the passage. The bath rooms are convenient, and well fitted with hot and cold water. The closets do not communicate with the wards. The walls look somewhat bare, and would be improved by the addition of a few pictures. The operation room is convenient, but requires rather more light. The mortuary is too small, and there are no conveniences for making *post-mortem* examinations.

All the duly qualified medical practitioners in the town constitute the medical staff, and the duty of house surgeon is taken in rotation. There is one nurse, with two servants and occasional night nurses under her, and the domestic arrangements are under the supervision of a committee of three ladies.

Rule VI. says—"Any proper patient, desirous of having the comforts and nursing at the hospital, may, on the certificate of a hospital surgeon, have the privilege of being admitted on payment in advance of not less than 20s. per week in the ordinary wards, or 30s. per week in a private ward." It is stated that the entire cost of working the hospital for the past ten years has averaged £1 5s. per patient per week.

Walker Hospital.—This hospital was built and furnished in 1870 by Mr. Mitchell, of the firm of Sir W. G. Armstrong, Mitchell & Co., but is now maintained by subscriptions from the employers and men of the several works in the district. It was established for " the reception of cases requiring surgical treatment, occurring in any of the large factories, ship yards, or other works in Walker, where the workmen contribute 6d. each per quarter towards the funds for the maintenance of the hospital." If sufficient accommodation exists, other cases are also admitted on payment of 10s. per week, and some may be admitted by letter, which, however, is only available for three weeks.

The management is vested in the hands of a committee composed of representatives of the subscribing works and delegates of the employers. This committee meets at the hospital every month. The staff consists of a surgeon, who visits daily and on the reception of accident cases, consulting surgeons, a nurse-matron, and an assistant nurse. There are on an average 75 in-patients and 250 out-patients attended to during the twelve months.

The building consists of:—*Ground Floor*—Sitting and bedroom for matron—lobby—front and back kitchens—operating room with bath-room—accident ward (5 beds)—out-door dressing and waiting rooms. *First Floor*—East ward (5 beds)—west ward (5 beds) —convalescent room with verandah—bedroom for nurse—store rooms—water-closet for patients—*Outer Premises*—Laundry in two rooms—mortuary—coal house and ash pit—wash-house—stick house—large yard—kitchen garden at back.

Ashburton and Buckfastleigh Cottage Hospital.— Open to all medical men; ten beds. One matron-nurse at a salary of £37 10s., with a servant under her at a salary of £12 a year. Extra help in nursing is occasionally required when there are very bad cases in the hospital. Water-closet on ground floor, adjoining male ward, but not opening to it. The premises are commodious, near the station, and with a large garden attached. The rooms are large, well lighted and airy, the entrance hall and landings capacious, and the staircase easy and con-

venient. The male ward is on the ground floor on the
left of the entrance hall. It is a bright cheerful room
with a bow window at the end, and a pleasant outlook on the green hills across the valley. There are
two beds, with room for three if necessary. All the beds
are Allen's patent. The washstands are enclosed,
stained, and varnished. The walls are coloured with
a warm French grey, and adorned with gifts of
pictures. On the opposite side of the passage is the
kitchen, spacious and well provided, and adjoining
this is a large scullery, with a boiler fitted for the
supply of warm water, which is an indispensable article
in hospital economy. Carrying out the cottage idea,
it is intended that the convalescents shall take their
meals, and sit in the kitchen. Over the male ward
and exactly like it, is the female ward (two beds).
Adjoining this, and over the general sanitary
accessories, which are on the ground floor, is the
surgery. Over the kitchen is the nurse's room.
A wardrobe or linen room opens out of this, and
beyond, over the scullery, is a large room in an
older part of the building, which has been converted into a ward with two beds. The landing
on this floor opens into a conservatory above the
porch. Then, on the second floor, are a committee
room, a single-bedded ward, and a store room. A
mortuary was erected in 1893. A general committee,
house committee, and a ladies' committee exist.
Recommendations are required except in cases of
accident or emergency, and a charge of 2s. 6d. per

week is made. Visits of friends are allowed three times a week.

DIET TABLE.

Ordinary Diet per Day.

Meat (cooked)	.	½ lb.
Potatoes .	.	1 ,,
Bread . .	.	1 ,,
Rice and Arrowroot .	.	2 oz.
Vegetables . .	.	,,

Ordinary Diet per Week.

Butter	. .	¼ lb.
Tea .	. .	2 oz.
Sugar	. .	½ lb.

Low Diet.

Beef Tea, Broths, Gruel, Arrowroot, Sago and Milk, as ordered.

West Cornwall Miners' Hospital, Redruth.—First established as a home for convalescent miners, with six beds, at the sole cost of Lord Robartes, and supported wholly by him. An allied society was afterwards started, having central and local committees, with the object of supporting additional patients. An arrangement was made, by which a certain number of beds (at the same time increased in number) was placed at the disposal of the society, on payment of an agreed sum per week per bed, when occupied,—Lord Robartes appointing, and paying the staff, and providing in every way for the maintenance of the patients and of the institution. On this basis it has ever since been worked. Up to 1871 it continued to be a convalescent hospital only, but in that year Lord Robartes built the accident wing. The subscribers pay 11s. per week for each bed filled in the convalescent side, and 13s. 6d. in the accident side. Lord Robartes reserves the right to fill beds himself, but does not now exercise it. All the expenses beyond these bed payments are borne by Lord Robartes. Patients are admitted on recommendation papers. The medical staff consists of the surgeons of the various mines, paid in the usual way, each looking after his own patients personally, or by substitute, and providing the medical necessaries required. Accidents are admitted without letters, and a hospital litter is kept. Female patients are not admitted, there being now a women's hospital in close proximity, joined by a corridor, but built and supported entirely by voluntary subscriptions.

The hospital stands on the high ground overlooking Redruth, and is surrounded by large and well-kept gardens, while the building, if plain, is of an attractive character. There are separate entrances and staircases for the convalescent and accident wings, which, however, communicate on each floor. Instead of large wards, the beds are grouped in rooms, four or five in each, and there are rooms which can be occupied by a single patient, if the nature of the case makes it desirable. On the convalescent side are spacious dining and sitting rooms ; the kitchen also is in this part of the building. Leading from the accident ward is an operating room, fitted with every requisite

of the most approved kind ; also a bedroom for the
occupation of the medical man in charge of a case,
should it be so serious as to require his attendance
during the night. The matron's room is on the convalescent side, that of the nurse on the accident side,
and everything is so arranged that immediate help is
always at hand. In serious accidents night watchers
are engaged in addition to the ordinary staff. The
bedding is kept in cupboards near the kitchen fire.
All the rooms are well ventilated, and that over the
kitchen, being naturally somewhat the warmest, is
generally appropriated to patients suffering from
bronchitis. Pictures and illuminated texts ornament
the walls of the rooms and staircases.

Petworth Cottage Hospital was built by the late Lord
Leconfield for the use of the town and neighbourhood
in the year 1868. It is endowed to the amount of
£33 8s. 6d. a year, yielded by a fund styled "watching fund," originally devoted to a different purpose,
as its name implies. It is supported by annual subscriptions, donations, payments by patients, and collections made in the church and chapel of Petworth,
as well as in the neighbouring parish churches, in the
form of " Harvest Thanksgiving." It is situated in
its own grounds, about half a mile out of the town,
and is about thirteen miles from the nearest general
hospital.

Description.—On the *Ground Floor* are—(1) Matron's room. (2) Convalescent ward. (3) Single-
bedded ward. (4) Kitchen. (5) Scullery. (*a*) Wash

and bake house. (*b*) Underground larder. (*c*) Pantry. (*d*) Wood-house, etc., with staircase to No. 10 ward for the conveyance of the dead.

On the *First Story* are—(6) Single-bedded ward. (7) Single-bedded ward. (8) Matron's bedroom. (9) Bath-room and water-closet. (10) Double-bedded ward.

And on the *Second Story* are—Double-bedded large attic. Single-bedded small attic.

A mortuary has been built since our last edition was published.

Patients pay from 1s. 6d. to 12s. per week each towards the expenses incurred. Usually this sum is paid for the patient by some charitable person, who stands at the same time as guarantor to the com-mittee. In the case of paupers it sometimes happens that they are sent in by private charity; sometimes that the guardians pay some sum, as 5s. per week, on the recommendation of their medical officer; and sometimes that all the expenses incurred are defrayed by the guardians. Convalescents are admitted in the summer or when there is room, at £1 1s. per week.

Wirksworth Cottage Hospital.—Much of the furni-ture was given. There is a nurse, who has occasional assistance when required. When her services are not needed at the hospital, she has to attend patients at their own homes during sickness on payment of the usual fees. Domestic arrangements are under a ladies' committee. There are two medical officers, but all medical men who have attended cases previous to their admission may continue the treatment if they desire.

15

Recommendations from subscribers are required. " Any out-patient, on recommendation of a subscriber and authorised by a medical certificate, will be provided with a dinner at the hospital at a small cost."

WIRKSWORTH COTTAGE HOSPITAL.

Three cottages were bought by a lady, who lets them to the committee at £8 per annum, and these were altered at a cost of about £120. The district extends to a radius of about five miles, with a population of 18,000, and the nearest infirmary is thirteen miles distant, *viz.*, at Derby. There are eight beds.

Reigate and Redhill Cottage Hospital.—This hospital has been enlarged by a gentleman in the neighbourhood so as to accommodate sixteen patients. It is limited to the district of the Reigate Poor-law Union, and is managed by a committee, with medical officer

and treasurer. "Any legally qualified medical
practitioner residing in the district shall be allowed

: REIGATE : COTTAGE : HOSPITAL :

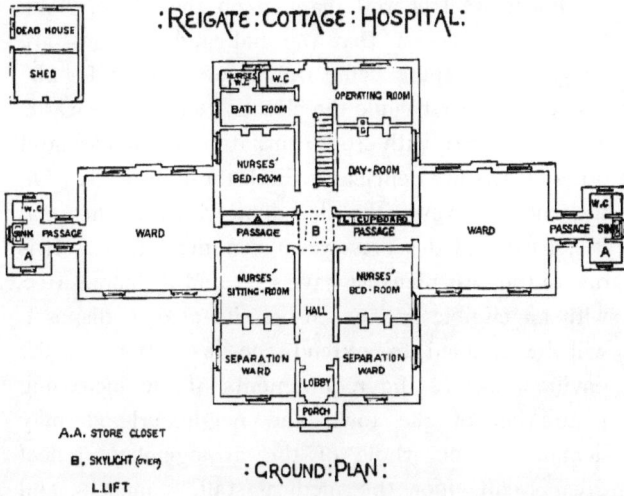

A.A. STORE CLOSET
B. SKYLIGHT (over)
L. LIFT

: GROUND : PLAN :

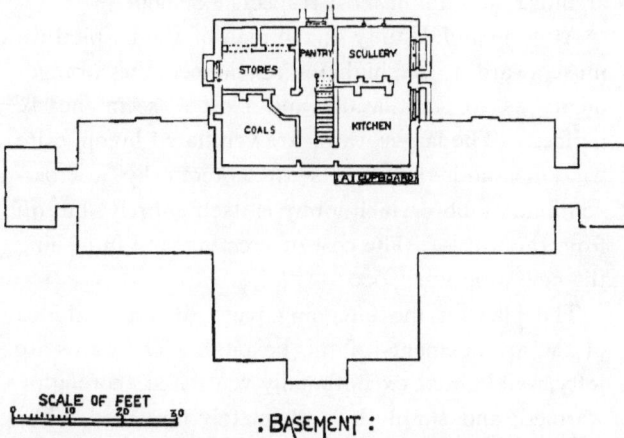

SCALE OF FEET

: BASEMENT :

to attend to his own cases in the hospital with the
consent of and in conjunction with the medical

officers." Subscribers of 10s. 6d. or donors of five guineas may recommend patients.

The hospital is well planned on the whole. We might suggest—1st, that the hall and passages are badly lighted, there being only one sky-light for the whole ; 2nd, that while the closets, etc., for the wards are well placed, with cross-ventilation, the nurses' and other closets are deficient in both these respects. On the whole, however, the plan upon which this hospital is constructed deserves warm commendation. It is one of the best plans we have seen, and the committee, with an eligible and extensive site at their disposal, will be enabled to extend the institution on the pavilion plan as the requirements of the increasing population of the town and neighbourhood may demand. The whole of the arrangements reflect great credit upon the medical staff, managers, and architect, and are in most respects excellent.

After a careful study of the plan of the hospital we must award it very high praise ; indeed, the arrangements as to ventilation and closets seem nearly perfect. The larger wards are ventilated by opposite windows, and the closets are entered by a cross-ventilated lobby, which lobby is itself entirely shut off from the wards. The cost of erecting and furnishing the new wing was £600.

The plan on the foregoing page gives a good idea of the arrangements of this hospital. The wards are lofty, well lighted, exceptionally ventilated, thoroughly warmed, and simply but adequately furnished. The water-closets are quite separate from the main building, being entered by a cross-ventilated lobby, which lobby is completely shut off from the wards. The building consists of one story, but two nurses' rooms have been placed in the roof. Should more hospital accommodation be wanted at any time, we would counsel the managers to build a second pavilion, and not to place wards above the present ones. The latter course would, for hygienic reasons, be best avoided. The one defect is, in our opinion, the situation and construction of the bath-room. Here two water-closets and a direct communication with the sewer have been brought into the interior of the hospital, and the precautions taken in reference to the ward water-closets are therefore partly neutralised. Why the nurses' water-closet should not be placed outside the building we fail to see, and the bath might with advantage be removed to the buildings where the ward water-closets are placed. Rufford's porcelain baths, not painted tin baths, should always be used for hospital purposes. The latter soon look dirty and become unsightly, whereas the former remain bright and clean for an indefinite period. The operation-room is exceptionally well lighted, and on the darkest days all available light will enter this room. We are informed that the hospital drains enter the town sewers; but all connection between the latter and the hospital buildings is cut by an open manhole, which is protected by a syphon placed between it and the main sewer, on the plan explained at page 191. Dr.

Walters has evidently taken great pains to make his hospital structurally complete, and we congratulate him upon a genuine success. It is true that a verandah cannot easily be connected with a hospital; but the open porch is not a bad substitute, and if the accommodation it provides proves to be insufficient for the increased number of patients, why not utilise the roof as an outdoor convalescent room? The view would be really charming, and the expense of the alterations could not be great.

The methodical way in which the in-patient register is kept is highly commendable. Its pages reveal an amount of good surgical work performed in an expeditious and thorough manner, which has naturally produced its reward—an exceptional success. A patient is admitted with spinal curvature, a Sayre's splint is applied, and the patient is made an out-patient on the same day; this occurs not once but often. An old woman of eighty, with a fracture of the left tibia and fibula, walked out of the hospital without assistance in less than two months. A case of epithelioma of the cervix was diagnosed, pronounced incurable, and the patient discharged in five days. Such are cases taken at random from the register, and they prove satisfactorily how crisply the Reigate Hospital is managed. It is far better administered medically than nine-tenths of the county infirmaries.

The situation of the hospital is all that can be desired; it has an ample garden, a mortuary, and a sedan chair for convalescents.

BOSTON HOSPITAL.

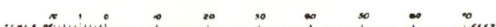

FIRST FLOOR PLAN

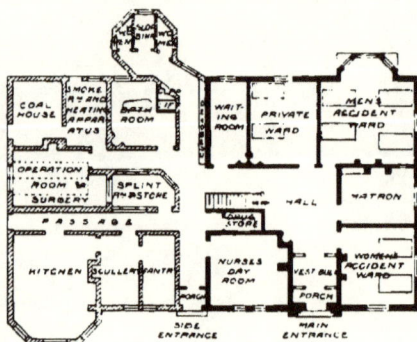

GROUND FLOOR PLAN

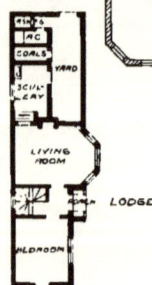

NOTE.—The new wing is shown by hatched lines.

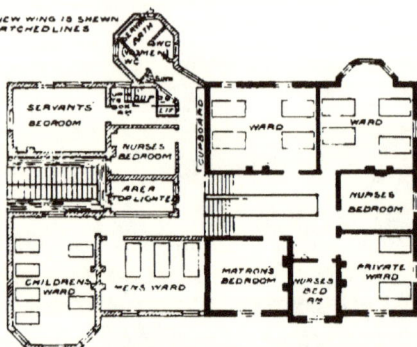

Boston Hospital.—An excellently planned and well arranged cottage hospital was opened at Boston, in Lincolnshire, in the year 1874.   Great credit is due to Dr. Mercer Adam, Mr. Pilcher, and the other members of the medical staff, who with the assistance of the committee and the honorary architect (Mr. Wheeler, to whom we are indebted for the facts contained in the following sketch), have managed to achieve a marked success in cottage hospital construction.   It will be seen that special care has been taken to give the wards an air of comfort and brightness, and we recommend any one who contemplates building a new cottage hospital to pay a visit of inspection to Grantham and to Boston, before finally fixing upon a plan.   We give the following detailed description, because there is much that is admirable in the fittings of the hospital :—

Description of the Boston Hospital.—In designing this building, two considerations were kept in view.   First, that it should be a *cottage hospital*, and therefore as nearly like a comfortable home as the circumstances of the case would permit, and unlike the large dreary-looking places formerly known as infirmaries ; and, secondly, that it should have a cheerful appearance throughout,—a great aid to the cure of sickness being gained by diverting the mind from the pain of the body with cheerful and pleasant surroundings.   A pleasant aspect, a beautiful look-out from the windows, large airy rooms with plenty of light, pictures with coloured texts and sentences on the walls of the wards, have all been brought to aid

the medical staff in their task of healing the sick.
The hospital is built with white bricks, relieved by
bands of moulded red bricks, having a handsome
moulded red brick cornice. The roof is covered with
blue Staffordshire tiles. Over the porch is a label,
made of encaustic tiles, with an effective border,
encircling the words " BOSTON HOSPITAL, 1874."
The front of the building faces due north, and on
this side are arranged the entrance porch, kitchen,
and other offices, the principal wards occupying
the south side. The building is entered by a
porch and lobby, with a double set of glass doors,
seats being placed between the two doors, so as
to provide a waiting place for visitors, or for the
convalescent patients to sit in on fine days. On
one side of the entrance, the motto " Rest and
be Thankful" is inserted in the brick wall, in encaustic tiles, and, to remove any encouragement to
a spirit of laziness which this might engender, on the
opposite side are the words " Work is Worship."
The lobby leads into a hall, out of which open
the downstairs wards, with the matron's room
in the centre. The hall is paved with red, buff,
and black tiles, and the walls to a height of
about four feet, with glazed bricks of a warm buff
colour, above which runs a string cornice of encaustic
tiles, with a pattern in black and buff, and above this
again are white bricks, finished under the ceiling with
an ornamental moulded red brick cornice. Under the
cornice is a band of black encaustic tiles, with buff

letters, containing the words, " Peace be to this house
and all that dwell therein," and on the opposite side,
" In God is our hope and our strength." In the
centre is a fire-place, the sides formed with chocolate
and buff picture tiles set into the brickwork. The
hearth is also formed with tiles ; but the whole hospital
is warmed by hot-water pipes. The name of each
room is placed over the door in tiles let into the wall.
Facing the porch is an alms-box with the words, " And
He saw a poor woman casting two mites into the
Treasury," and over it is inscribed the sentence from
Proverbs, " Say not unto thy neighbour, Go and come
again, and to-morrow I will give, when thou hast it by
thee." The wards on the south side look over the
public recreation ground. All the windows throughout the building are made in three parts,—the two
lower consisting of the ordinary sashes, made to lift
up and down, and over these is placed a sash a foot
deep, hung on hinges, so as to fall forward into the
room. The air, being thus directed upwards towards
the ceiling, does not fall directly on the persons
occupying the room. A similar arrangement is provided over the doors, so that an uninterrupted current
of air may pass through the room without creating a
draught. Provision is also made for the escape of the
foul air from the rooms by means of ventilating flues,
which pass up by the side of the smoke flue openings,
covered with fine perforated zinc, which is let into
them immediately under the ceiling. The air in these
shafts is rarefied by the heat derived from the smoke

flues which adjoin them, and consequently there
is always an upward current drawing the vitiated
air from the wards. Escape is provided just above
the ridge of the roof; the air flues terminating by
projections from the chimneys, which have the appearance of small buttresses supporting the chimney
stack, and while fulfilling this useful purpose, add
greatly to the ornamental character of the chimneys.
The walls of all the wards are painted in cheerful
colours with silicated enamel, which is quite hard and
impervious, and can be washed down frequently. In
some wards the lower part is painted for a height of
three feet six inches, the separation from the dado being
made by a moulded rail, the two smaller members of
the moulding being coloured respectively black and
red. These coloured lines afford an agreeable con-
trast to the colour of the walls, and give a cheerful appearance to the room. Immediately under the
ceiling, and running all round the wards, are letters in
Old English character, coloured vermilion, the capitals
of which are blue. The following are the stated
sentences :—" Despise not thou the chastening of the
Lord, nor faint when thou art rebuked of Him : for
whom the Lord loveth He chasteneth, and scourgeth
every son whom He receiveth." " Keep innocency,
and take heed unto the thing that is right, for that
shall bring a man peace at the last." " As thy day is,
so shall thy strength be." " Contentment with Godliness is great gain." The fire-place is formed by a
neat cast-iron curb, to which are attached the bars of

the grate, and round this is a double row of blue
picture tiles, and the whole is finished with a wooden
moulding, stained and varnished. In one ward the
cross mouldings are double, leaving a space between,
in which are cut the following letters, picked out with
vermilion, so that they have a very bright effect :—
" In God is my health and my glory." The hearth is
also composed of similar tiles to the front of the grate,
and finished with a neat iron curb, which acts as a
fender. The cost of the whole grate, with tiles and
fender complete, is not more than that of an ordinary
grate and plain mantel-shelf suitable for such a ward.
The grates in all the wards are similar, the subjects
of the tiles being varied by descriptive designs of
fables, flowers, etc. On the walls are hung coloured
pictures in simple frames, selected from those issued
with *The Illustrated London News*, *The Graphic*, etc.,
and those supplied by the Tract Society.

The surgery is fitted with cupboards and shelves, a
sink and water taps. The water-closets, etc., both
downstairs and on the upper floor, are detached from
the building and approached by separate passages.
On the opposite side are the kitchen, scullery, and
pantry. The laundry, wash-house, drying closet, etc.,
are in separate out-buildings, near which there is
also a mortuary with a separate entrance. The
staircase, leading to the floor above, is of pitch
pine varnished, the chamfers, etc., being picked
out with black. The treads are broad and easy,
and the width sufficient to allow a stretcher,

containing a patient, to be carried up and down. Communicating with the landing are lavatories and separate water-closets for the use of male and female patients. From this landing the stairs divide into two separate flights, each leading to a gallery on either side of the building, by which access is obtained to the upper wards,—a separation being thus effected between the male and female wards. The windows are so arranged that the patients lying in bed can see over the recreation ground, and enjoy a beautiful and cheerful view. The staircase is open to the roof and covered in by a lantern, glazed at the sides with coloured cathedral glass with small perforated zinc openings, allowing a current of air to circulate throughout the building and into the wards by means of the fanlights over the doors.

In the Jubilee year subscriptions were obtained for enlarging the hospital, and a new wing was added, in which the kitchen and offices were placed on the ground floor, and separated, as far as practicable, from the rest of the hospital. The surgery and operation room were also removed to this wing, and provided with light from the roof. Over the kitchen and offices two large wards were added, and additional rooms were also provided, over another part, for nurses' and servants' bedrooms. A hoist was at the same time provided for lifting coal and other heavy articles to the upper floor. A detached laundry has been erected, the whole of the walls of which are lined with glazed bricks, and

every convenience has been provided for washing the
clothes and linen.

The hospital now contains 30 beds, including a
very beautiful children's ward, with six cots, and all
the appliances needed for the little patients. The
staff consists of a matron, a staff-nurse and five
lady-probationers, who come for one or two years'
training, and pay 10s. a week, and 1s. for laundry
expenses. The servants consist of a cook, a ward-
maid, a housemaid, and a laundry-woman.

The building stands upon about half an acre
of ground, leased from the Corporation of Boston for
ninety-nine years at a nominal rent. The space not
occupied by the building is laid out as a garden for the
use of the patients. In the grounds near the entrance
gate, a lodge, where a man and his wife reside, has
been erected at the expense of a gentleman in the
neighbourhood as a memorial to his brother. The
man, in consideration of the house-room, coals, and
gas, acts as porter, keeps the garden in order, and
assists when required in the hospital. There are
also two bedrooms in the lodge for the use of
nurses.

The hospital and wing added in 1888 were built
from the plans and under the superintendence of
Mr. W. H. Wheeler, of Boston, the honorary
architect. The total cost of the original building,
including the fencing and arrangements for the
garden, laying on the water, providing a portion of
the furniture, and every other expense, has been

£2165. This is exclusive of the cost of the porter's lodge and the furniture for four of the wards, which were furnished by four families.

The number of patients treated during the year ended 30th June, 1895, was 203. The patients pay 5s. a week, and children half-price. Private patients are admitted on payment of not less than £1 per week, which does not include medical attendance or stimulants.

Yeatman Hospital, Sherborne.—At the death of the Rev. H. F. Yeatman, of Stock House, Sherborne, the magistrates of the county decided on establishing a permanent testimonial to his memory, and subscribed £500 for that purpose, which was supplemented by £500 from the inhabitants of Sherborne and the neighbourhood. A joint committee was appointed to carry out the wishes of the subscribers, and at a meeting held in Sherborne it was unanimously agreed that a hospital in the town would be a fit memorial. Plans by the late Mr. Slater were adopted, the centre and west wing being first erected, and opened for the reception of patients on the 19th March, 1866. In presenting their first annual report the committee appealed to the public for funds to enable them to complete the building, and in answer the ladies of Sherborne and the neighbourhood held a bazaar in the Town Hall, which realised the handsome sum of £557 13s. 6d. This, with other donations, enabled the building to be completed at a cost of upwards of £4000, the accommodation consisting of twenty-four beds.

Braintree and Bocking Cottage Hospital.—This little hospital, which is pleasantly situated in the neighbourhood of Halstead, Essex, was founded, and for a long time maintained, by the late Mrs. George Courtauld. No one who had the privilege of knowing Mrs. Courtauld will need to be reminded of her worth. Charitable to a high degree, she yet tempered her gifts with wise discretion. No mere distributor of doles, she was, in fact, most justly known as the friend of the poor. The universal regret caused by her sudden death is shared by the writer of this book, as he knew no one whose judgment and counsel he more thoroughly trusted. The Braintree Hospital is a proof of her worth. Unaided she drew up the rules and regulations for its management : and those rules are the embodiment of the best principles of cottage hospital management. The letter of recommendation is the best the author has seen, and he reproduces the footnote for general adoption. It could not be better, and it testifies to the sound principles upon which Mrs. George Courtauld conducted her little hospital :—

"To avoid disappointment and needless trouble, it is desirable to state that the hospital is not intended to be used as a sick home for the reception of indigent sick people, but as a hospital for the treatment of cases where cure or permanent benefit can be reasonably anticipated.

"The nurse is strictly forbidden to receive money from the patients."

The Braintree Hospital has been carried on since

16

1885 by a committee, and the work so wisely begun
has been well continued.

Chronic Hospitals.—It has of late been often urged
by the medical press, and in other quarters, that
infirmaries or hospitals for the sole treatment of
chronic and incurable cases would meet a felt want.
For our own part we must state our belief that whilst
the establishment of such institutions may be desirable,
still it can never be defensible except each person is
made to pay fairly for the advantages and treatment he
receives. Miss Black, of South Kensington, humanely
founded a chronic hospital for the poor of Hampshire
in 1872, at Northam, Southampton, to which she devoted the whole of her energies. Miss Black has
exercised great self-denial, and the following description of the beneficent work she has so successfully
carried out may prove interesting to many readers.
Writing from 14 Longridge Road in 1880, she said :—
" I found your paper on arrival at my little hospital
yesterday, and hasten to answer your questions. I
began, continued, and carried it on all by myself. At
first I had two beds ; but finding it absorbed too much
money I gradually admitted only those who could
come daily, and year by year the numbers rapidly
increased, and now they come from all parts of
Hampshire. My hospital is solely for the treatment
of ulcered legs and eczemas. I have got the money
entirely by working hard at raffles, entertainments,
etc. I have a matron and now four nurses, the former
resident, all of my own training. Such patients as

are at first too ill to attend, I send my head nurse to, and she daily reports progress. Now that we live in London I go down every week and sleep there, and on Wednesday I see and dress from forty to eighty cases. I have had ninety and a hundred and five in one morning."

This letter shows how much one person with determination and energy can do to relieve suffering humanity. Would that each town and county had a little institution to minister to the necessities of the suffering poor!

Saffron Walden.—Two rules deserve reproduction and imitation :—

1. No person shall be admitted unless able to maintain himself or to pay for his cure, unless he be suffering from severe accident. and not then if he can be efficiently treated at his own home. In case any such person is accidentally admitted he shall be held accountable to the committee for all expenses incurred on his behalf.

2. All unmarried patients who are members of a club shall pay to the hospital towards their maintenance two-thirds of the weekly allowance they receive therefrom.

At this hospital trusses are supplied for a nominal payment to persons known to subscribers. This is a good rule, and would occasionally be an immense boon to the poor.

CHAPTER X.

SELECTED AND MODEL PLANS CRITICISED AND
COMPARED, WITH A DETAILED DESCRIPTION
OF VARIOUS HOSPITALS.

Description and criticism of plans of Cottage Hospitals at
Grantham, Maidenhead, Petersfield, Ashford, Stamford
(Fever), Bourton-on-the-Water, Beccles, High Wycombe,
Cheshunt, Milford (Staffs), Forres Leanchoil, Watford,
Mirfield, Falmouth, Brixham, Dartford, Wood Green, and
Surbiton—Plans for model Pavilion Hospitals of various
kinds—Small Hospital with nine or twelve beds—General
Hospital with thirty or forty-eight beds—Convalescent
institutions—Temporary and permanent Fever Hospitals
—Isolation Hospitals, Local Government Board's model
plans—The construction of small Cottage Hospitals.

IN Chapter VIII. we have gone into much general
detail when dealing with the questions of cottage
hospital construction and sanitary arrangements,
and it will not be necessary for us to repeat our
views on these questions here. At the same time,
it was pointed out in reviews of the first edition that
very few plans were given, and that a brief descriptive
chapter devoted to this subject alone would be of
real service to many. It has been determined, therefore, to accede to the wish thus expressed, and the

GRANTHAM HOSPITAL

LAUNDRY
BLOCK

ISOLATION BLOCK
FOUR BEDROOMS ON 1ST FLOOR
TO PORTERS HOUSE

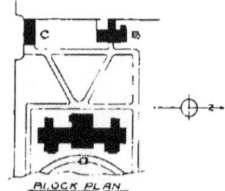

BLOCK PLAN
A. HOSPITAL
B. ISOLATION BLOCK
C. LAUNDRY

GROUND FLOOR PLAN

FIRST FLOOR PLAN
TWO BEDROOMS FOR SERVANTS
AND BOX ROOM ON 2ND FLOOR

[To face p. 245.

plans about to be described have been carefully selected from a large available number kindly placed at our disposal. No plan has been omitted on the ground of expense. Far from it : all have been carefully considered on their merits, and the following have passed the scrutiny with more or less credit.

First and foremost, quite a head and shoulders in advance of its fellows, and probably, hygienically speaking, the most complete in the country, stands out the *Grantham Hospital, Lincolnshire.* Fortunately the committee had very clear views upon the subject of sanitary arrangements. They very carefully altered, amended, and rearranged the plans first submitted by the architect, until the result was in this respect satisfactory. Twenty years ago, when the hospital was built, it was considered a model of hygienic perfection, and since then the committee have from time to time made such alterations as have been necessary in the light of modern improvements. It may be mentioned that all the suggestions as to improvement made in our second edition have been carried out. The hospital is built of stone, and being situated upon the side of a hill, and commanding an excellent view, it is certainly a delightful place for an invalid. Indeed, when looking out of the board and nurses' sitting room one instinctively wishes that if struck down by illness such excellent accommodation might be within easy reach. The plan of the site is excellent, and the arrangements and shape of the wards novel, pleasing, and noteworthy. They present a cheerful and airy appearance which fills the visitor with pleasure. Much has been said about the advantages of octagon and circular wards ; and here at Grantham is another plan which merits imitation. We commend it to the attention of architects generally. The laundry, mortuary, and isolation ward are well and thoughtfully placed, and will bear careful inspection. Since the second edition was published an airy children's ward has been erected, containing ten cots. A special feature is the new kitchen, built of stone and lined with glazed white bricks, and being fitted with lantern ventilation no smell of cooking can penetrate to the wards. New closets have also been built, separated from the main building by lobbies. But the points which reflect the highest credit upon those who manage this hospital are its drainage and sanitary arrangements. It would be, perhaps, too much to say that any system of drainage was perfect, but we may fearlessly state that every modern improvement and precaution is to be found at Grantham. The drains are placed entirely outside the buildings ; they are made of four, six and nine-inch sanitary pipes jointed with cement. The main drain joins a nine-inch sewer on the public road : this is freely ventilated by direct openings to the external air at frequent intervals. A manhole is placed at the head of this sewer, and in it there is a flushing valve ; when this is let down it admits of the drain being filled with water and flushed out. The main drain from the hospital is disconnected

from the manhole by a syphon trap, on the hospital
side of which is an opening from the drain to the external air, protected by an iron grating on the surface
of the ground. At the head of each branch drain a
light three-inch pipe is carried up the side of the
buildings, with as few bends as possible, two of them
to the ridge of the roof, and left open at the top. A
four-inch lead soil-pipe in the main building, and another in the isolation ward, are carried up full bore
above the roofs and left open at the top. A direct
communication is thus established between the opening in the drain on the hospital side of the syphon
trap at the manhole, and the open ends of these pipes,
so that the drains are freely ventilated and filled with
fresh air, all sewer air being excluded from them by
the syphon trap. Places are provided at the heads
of the drains where water can be run in to flush
them.

All bath, lavatory, sink, and other waste pipes end
in the external air over gulley traps, so that there is
everywhere complete air disconnection, and no direct
communication between the drains and the interior of
the hospital.

The rain water from the roof of the main building
is conveyed through sanitary pipes, laid at a less
depth than the drains, to an underground tank beneath
the wash-house ; a gulley trap is sunk low enough to
receive the overflow from this, so that it is open
to the external air, and disconnected from the
drain. The rain water from the isolation ward and

the mortuary and *post-mortem* room is not collected, in case it might carry infection, but it is run into the drains, the down spouts ending over gulley traps.

The closet apparatus are the patent wash-down Tornado, with adjustable rising seats, and connected with a large syphon-flushing cistern and service pipes. The hoppers, or slop-sinks, are large, round, enamelled earthenware, with three-inch flushing rim, and supplied by a forty-gallon flushing cistern to each, with three-inch supply pipe, and connected with a four-inch patent lead trap, which discharges into a gulley trap outside the buildings.

The hot water for heating is provided by a powerful independent cylinder boiler fixed in the basement, and supplying a 200-gallon cylinder, from which the supply is carried to the whole of the baths, lavatories, and sinks. The hot water for the isolation wards is supplied by a patent gas geyser.

The Grantham Hospital cost £5364 to build, £812 to furnish ; total, £6176. Since when £297 has been spent in new floors, £605 in building children's ward, £120 in new drainage works, £836 for the new kitchen, nurses' pantry, and enlargement of matron's room, £200 for the new lavatories, and £178 for the new bath room and lavatories to isolation ward ; total, £8412.

Maidenhead Cottage Hospital.—The plan of this hospital is generally good, compact, and economical. The building is arranged on the pavilion system,

and contains accommodation for eight patients. There
are two wards for three beds each, and two for one

MAIDENHEAD COTTAGE HOSPITAL.

bed each. The wards are so situated that their
principal windows face the south, and provision is
made for thorough cross ventilation by means of the

doors and windows. At the end of each of the general wards are the lavatories, clothes' cupboards, etc., which are separated from the building by lobbies for ventilation. Near to and overlooking the wards are the nurses' bedrooms. The surgery and the committee room open out of the entrance hall.

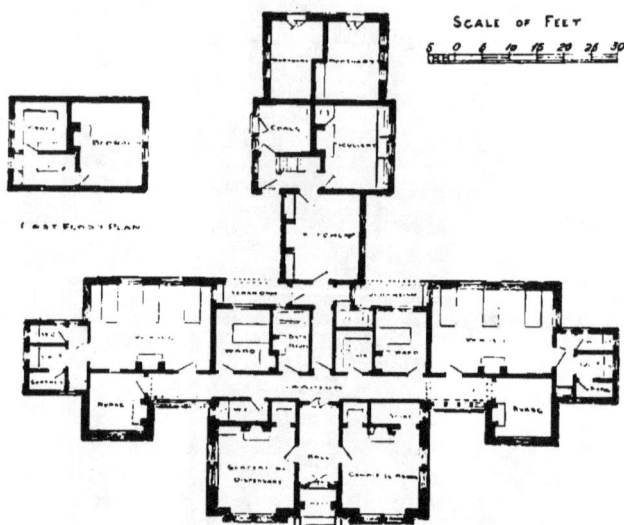

MAIDENHEAD COTTAGE HOSPITAL.

A bathroom and a linen room are also provided. The plan is so arranged, however, that the latter will be removed when the additional story is added to the central block, so as to make room for the staircase, as shown by dotted lines on the plan. It will then be placed on the one-pair story.

The kitchen is situated at the back, and has a lobby of separation in order to prevent as much as possible the smell of the cooking having access to the main block. A scullery and mortuary have been added on the south of kitchen. The latter is properly fitted, and supplied with water.

The buildings are faced with red and grey bricks, with Bath stone lintels and sills, and the roof is covered with Broseley tiles. The general effect thus produced is very pleasing.

The water-closet, which opens directly into the centre corridor, is wrongly placed, but it is stated that it is rarely used. The kitchen is excellently situated for a hospital of ten to twelve beds. This plan offers many advantages. It reflects great credit on the architect, and the hospital will well repay a visit. It cost £1925 to build, £350 for furniture ; making a total outlay of £2275.

Petersfield Cottage Hospital.[1]—The situation of this hospital is probably more charming than that of any other in Hampshire. The view from its windows in the summer is really lovely, and the air quite a treat to breathe. The wards are well lighted, well ventilated, and airy ; the new operation theatre is excellent in situation, and the lighting and fittings are all that can be desired. Unfortunately, the architect has, as usual,

---

[1] The plan and description which follow are what appeared in the second edition, no reply having been received to our inquiries.

marred an otherwise admirable plan by his total disregard of sanitary arrangements. Every closet is so

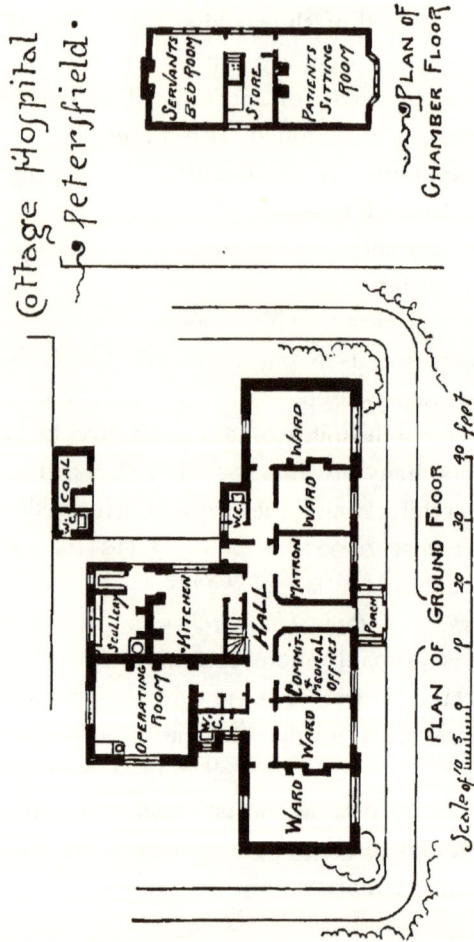

Cottage Hospital, Petersfield.

PLAN OF CHAMBER FLOOR

PLAN OF GROUND FLOOR

placed that disagreeable smells perceptibly arise from

PETERSFIELD COTTAGE HOSPITAL.

ASHFORD COTTAGE HOSPITAL.

it, and often fill the corridor and small wards. This is
not creditable, especially as the medical staff have
done their best to get the evil remedied. There is
plenty of room to build out these conveniences in
each case, and to separate them from the hospital
by a cross-ventilated lobby. A bathroom would be
an advantage. Pleasantly situated, admirably conducted, and possessing accommodation for from eight to
ten patients, this hospital is also well worthy of inspection. The arrangements for heating the hall, and the
garden plan are particularly good. It originally cost
£1100 to build, and £234 to furnish; total £1334. In
1879 the operation theatre was added at an additional
outlay of £300.

Ashford Cottage Hospital.—The elevation of this

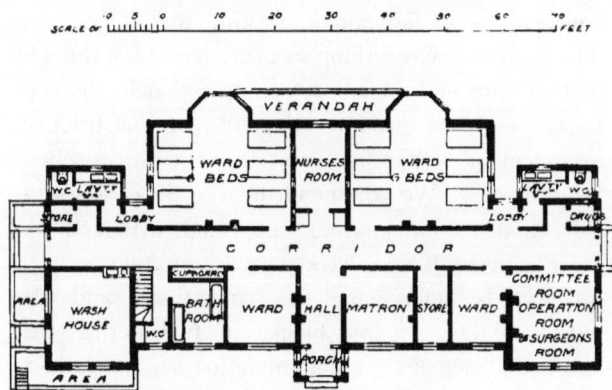

ASHFORD COTTAGE HOSPITAL.

GROUND FLOOR PLAN

hospital is tasty and effective. The ground plan is
novel, and possesses many admirable features; *e.g.*,

the verandah, the bay windows to the large wards, and the operation theatre. It is a pity, however, that the wash-house has been attached to the hospital proper. The position of the water-closets, lavatories, and sinks was altered in 1894 at a cost of £177, and these offices are now placed outside the building and cut off from the corridor. The kitchen is below the wash-house in the basement. The operation theatre serves as a surgeon's room, and also for committee meetings. The hospital contains fourteen beds. It cost £3140 to build, £200 to furnish, and £177 to alter as mentioned above; making a total of £3517.

The Fever Wards, Stamford.—We include a description of these wards here, because, although attached to a general hospital, they are built as separate cottages or blocks, and have much to commend them. The description given by Dr. Newman in his history of the Stamford, Rutland, and General Infirmary is so full and interesting that it is reproduced below. Everything seems to have been thought of that care and ingenuity combined could discover to be likely to promote the efficient construction, arrangement, and administration of these excellent fever blocks. We commend them to the candid consideration of sanitary authorities throughout the country.

"The fever blocks have a space of from forty to fifty yards intervening between them and the older building. Three blocks of two stories each afford on each floor accommodation for five patients and a nurse. These blocks are practically uniform in arrangement, the central block differing, however, somewhat in external appearance and in the internal plan from the other two ; a distance of from seven to ten yards exists between the several blocks. Behind these structures is an open space, fifty yards deep, left purposely for the erection at some future time, if it be found desirable, of a kitchen, or other offices. At the further end of this space is placed a low one-storied building containing the appliances for a laundry and disinfecting chamber, while in the rear of this again is the dead-house, also quite detached. Especial care was taken to cover all the enclosed area with a layer of Portland cement concrete, six inches thick, so as to prevent the ascent of damp from the subjacent porous oolitic strata. A damp-proof course was also provided at the base of the walls. The buildings are of the local stone of the district, oolitic, of varying density. The quoins, jambs, and window-beads are of Casterton stone, the window-sills and plinths of Clipsham limestone.

" Each block has two stories above the ground-level, with a cellar in each basement. Every floor is thus arranged :—There is an entrance-lobby, having on one side the stone staircase, and on the other side a nurse's room, while between the two is placed the door of the ward, which, with its appendages, occupies the remaining space.

" The wards are twenty-five feet square, with a height of fifteen feet, and are arranged to receive five patients : the air-space for each bed, therefore, be ng

17

over 1800 cubic feet. The walls are lined with
glazed bricks throughout, built in as the work pro-
ceeded, and jointed in Parian cement. The windows
are placed on the three outer sides of the ward ;
two, smaller in size, face the door of entrance,
while the larger windows on the right and left hand
respectively are directly opposite to each other. In
their lower two-thirds the windows are of the
ordinary sash-pattern, while the upper third is occupied by a framed casement, which is hinged at the
bottom and falls inwards at pleasure. This arrangement has been adopted in all the windows. Each
ward has a bath-room and water-closet, opening,
with the intervention of a cross-ventilated lobby,
from the corners most distant from the entrance
door. This description applies to the two end blocks
only ; in the central structure the bath-room and
water-closet are on the right and left of the entrance door in the corners of the return walls.
These additions are lined with Parian cement ;
the staircase, nurse's room, etc., with ordinary
plaster. Four rectangular metal ventilating shafts
for the supply of fresh air are fitted in each ward,
opening below the floor level to the external air,
and ending within the ward about five feet above the
floor.

" A central stove on Galton's pattern is placed in
the ward ; to this fresh air is brought by special
wide channels from the outside, and this air, warmed
in its passages, is delivered into the room through

perforated openings above the stove, three feet from
the floor. The smoke-flue is continued straight upwards to the open air, running within a square
framing of metal-work covered with tiles. Not quite
one-half of the sectional area of this framing is,
however, occupied by the smoke-flues; the remaining
portion is so contrived as to form a ventilating shaft
for the two wards, extracting by special gratings
the foul and heated air close under the ceiling, and
then delivering it through openings of proportionate
size placed on the sides of the chimney some
distance above the roof. The entrances and the
nurses' rooms are paved with hard, well-burned red
and black Staffordshire tiles. The ward floor is of
hard pitch pine, long dried before use, and closely
joined by grooved and tongued joints, these floors
being made, by a coating of paraffin or varnish,
impermeable to moisture. Under all these wooden
floors there is provision for the free circulation of air
between the concrete below and the joints above.
The whole of the internal woodwork has been so
arranged that there are no mouldings or projections
round the doors or windows; no facilities, in short,
for the collection of dust. The woodwork, again,
internally is all varnished, not painted.

"The glazed linings of the ward have been thus
planned:—Next the floor are placed two courses of
black bricks; then a dado, about four feet high, of
cream-coloured brick, finished above by a single band
of chocolate colour; from this darker line to the ceil-

ing the wall is covered with bricks of greyish white.
To break the uniformity and coldness of this colour,
three tile pictures (three feet by two feet) have been
placed, one on the inner face of each external wall.
These pictures illustrate in each instance some agricultural or outdoor occupation, and are let in flush
with the inner surface of the wall itself. In the lobby
of the bath-room is fitted a plain slate lavatory. The
baths in most of the wards will be on wheels, so as to
allow of easy movement to the bedside of any inmate.
The closets are throughout on Jennings' trapless
pattern ; they open directly without syphon or bend
into an earthenware soil pipe, constructed of jointed
lengths of glazed sanitary pipes, specially made for
the purpose. The upper end of the soil pipe open to
the external air, is guarded by a Field's cowl, and the
channel, fixed to the outside of the wall, discharges
below into a main sewer, with the intervention, close
outside the wall, of a Pott's Edinburgh trap. All the
waste pipes from closets, cisterns, or baths, from sinks
in the nurses' rooms, or lavatories, are so constructed
as to ensure perfect disconnection. In each instance
they are carried through the outer walls, run down the
outside, and open upon hollowed stones a foot or
more distant from a trapped iron grating lying upon
the ground-level : sewer gas, if it should regurgitate
through the traps, cannot therefore ascend through
these pipes into the building. The branches of sewer
from each closet, bath-room, and surface-trap are
collected into one large channel, which leads down to

a closed but yet well-ventilated cesspool. This is fitted with convenient arrangement on the ground-level for such frequent emptying as may be desired. In the line of the main sewer, 140 yards in length, there are two or more apertures for the escape of sewer gas. The disinfecting apparatus which has been selected is on the pattern suggested by Dr. Ransome of Nottingham.

"The whole arrangements have been carefully worked out by Mr. Browning, the architect of Stamford. The exterior is extremely plain, with no elaborate ornamentation, and the internal arrangements at least allow the hope that perfect isolation, with every convenience for the sick inmates, will have been attained." The cost of the three blocks, including internal fittings, furnishing, and all other expenses, was £7500.

Bourton-on-the-Water.—This hospital, which is plainly and inexpensively built, occupies one of the best sites in the village. It was erected in 1878, and opened in the autumn of 1879. It has accommodation for eleven patients. The wards are well proportioned, light and airy, those on the first floor being especially pleasant. In the second edition of this book we noticed that the sanitary arrangements of this hospital were decidedly unsatisfactory. The surgeon, Mr. Moore, raised a special fund to place these matters right, and a complete system of drainage has been established. The building cost £1100, the furniture £180; total, £1280.

Beccles Hospital.—An inspection of this hospital is

well worth the trouble. There is something to see and learn there ; and the general arrangements are on the whole good. On the right on entering the hospital is the surgeon's room, with waiting room and dispensary attached. On the other side are arranged the matron's room, kitchen, storeroom, scullery, etc. ; while opening out of the central hall is the accident ward with two beds. On the first floor are the male and female wards, each containing five beds and one cot ; the Jubilee ward with three beds ; the operation room ; a private ward ; and the matron's bedroom. It would have been better if the operation room had been on the ground floor near to the accident ward.

In the second edition we mentioned that the water-closets on the ground and first floors were scarcely satisfactory, and that a bath-room was much needed. We are very pleased to find that our suggestions have been accepted, and that additions and alterations have been made which have given the hospital a new ward with three beds and a bath-room, and that the water-closets are now practically disconnected from the hospital. All the wards are well proportioned, airy, and cheerful ; the accommodation is ample, and the plan of the garden and elevation most satisfactory. There is accommodation for eighteen in-patients, and the out-patient arrangements are well planned and satisfactory.

High Wycombe, and Earl of Beaconsfield Memorial Cottage Hospital, Buckinghamshire.—This is a novel plan. Advantage has been taken of an inequality in the site to place the kitchen and offices in the basement with

BECCLES HOSPITAL.

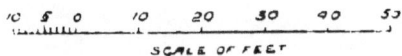

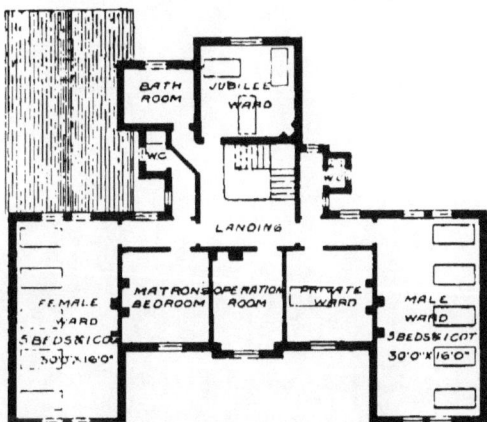

SCALE OF FEET

10 5 0   10   20   30   40   50

FIRST FLOOR PLAN

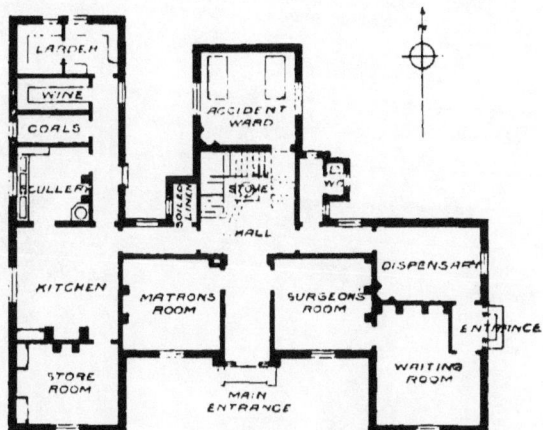

BATH ROOM

JUBILEE WARD

WC

WC

LANDING

FEMALE WARD 5 BEDS & 1 COT 30'0" × 16'0"

MATRONS BEDROOM

OPERATION ROOM

PRIVATE WARD

MALE WARD 5 BEDS & 1 COT 30'0" × 16'0"

GROUND FLOOR PLAN

LARDER

WINE

COALS

SCULLERY

KITCHEN

STORE ROOM

ACCIDENT WARD

SOILED LINEN

STORE

WC

HALL

MATRONS ROOM

SURGEONS ROOM

DISPENSARY

ENTRANCE

WAITING ROOM

MAIN ENTRANCE

a mortuary at the back of all. The situation is pleasant, the garden is well laid out, the elevation is homely
and well adapted to the purpose of a cottage hospital,
and the general arrangements are good. The water-
closets and lavatories are, however, insufficient when
the hospital is full, and are placed at an inconvenient
distance from the wards. An operation and con-

COTTAGE HOSPITAL. HIGH WYCOMBE.

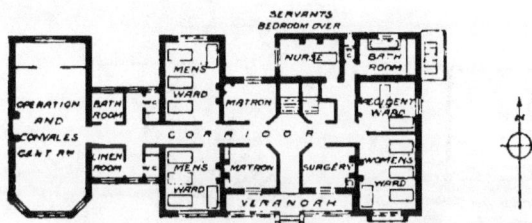

GROUND FLOOR PLAN

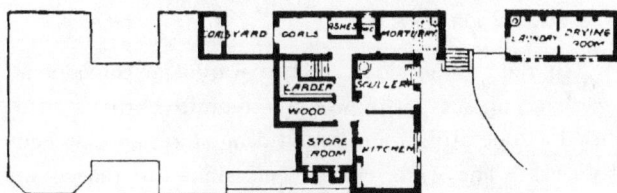

BASEMENT PLAN

valescent room has been provided. There are
many good points, the provision of a mortuary not
the least; and the architect undoubtedly deserves
credit, on the whole, for the satisfactory plan he has
prepared. For a rural district we have no doubt it
answers well, though in a mining or colliery neighbour-

hood it would be difficult to administer, and would require re-arrangement in some points. The original cost for building was £1240, furniture £260; total, £1500. There is accommodation for fourteen patients.

Cottage Hospital, Cheshunt, Herts.—This plan is given as an example of a small, economically planned hospital, perhaps more strictly entitled to the name " Cottage " than some others of equal size. It contains a ground story and two rooms formed in the high-pitched roof of the central portion.

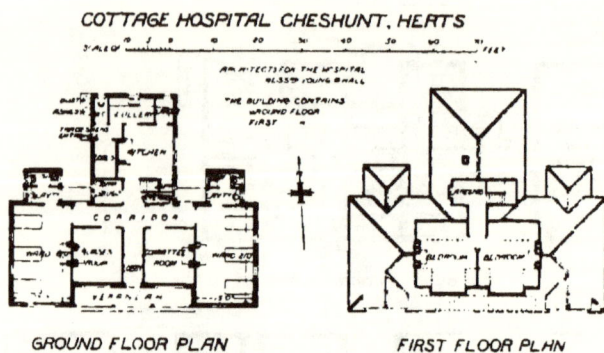

COTTAGE HOSPITAL CHESHUNT, HERTS

GROUND FLOOR PLAN   FIRST FLOOR PLAN

On the ground floor are two wards for three beds each, a nurses' sitting-room, committee room, also used as operating room, small drug store, and kitchen offices. The water-closets and sinks are placed in small wings projecting to the north, and the connecting lobbies serve both as lavatories and for access to the garden.

The building was designed by Messrs. Keith Young & Hall, and cost £1010.

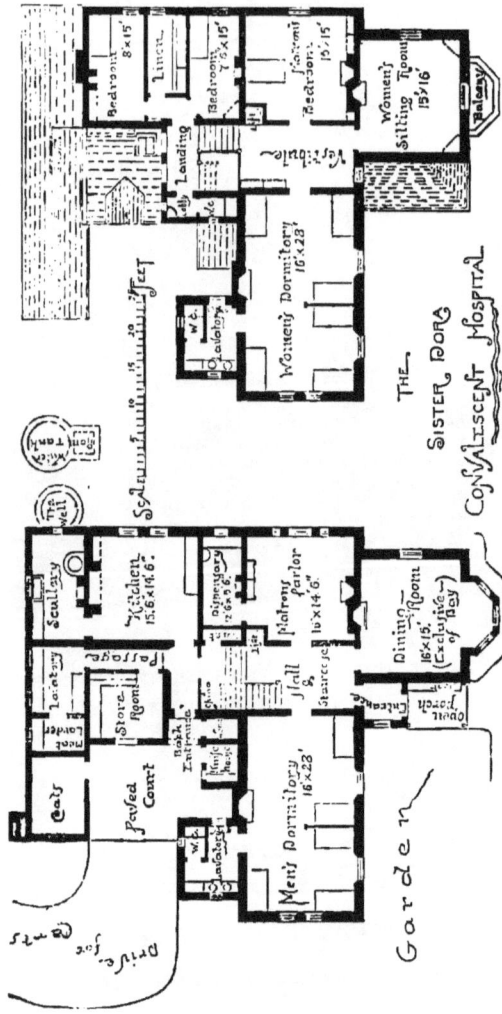

PLAN OF UPPER FLOOR

Bedroom 8'×15'

Linen

Bedroom 15'×15'

Bedroom 15'×15'

Women's Bedroom 15'×15'

Women's Sitting Room 15'×16'

Balcony

Landing

Vestibule

Women's Dormitory 16'×23'

W.C.

Lavatory

THE SISTER DORA CONVALESCENT HOSPITAL

Scale in Feet

Hot Water Tank

The Well

Scullery

Kitchen 15'6"×14'6"

Dispensary 12'6"×9'6"

Matron's Parlor 16'×14'6"

Dining Room 16'×15' (Exclusive of Bay)

Passage

Store Room

Meat Larder

Back Entrance

Coals

Paved Court

W.C.

Lavatory

Hall & Staircase

Open Porch

Entrance

Men's Dormitory 16'×23'

Garden

Bridge for Carts

PLAN OF GROUND FLOOR

The High Road

*The Sister Dora Convalescent Hospital, Milford,
Stafford.*—This hospital was erected mainly through
the exertions of Miss Margaret Lonsdale as a
memorial to Sister Dora of Walsall; the whole of
the profits resulting from the sale of the biography,
Sister Dora, having been devoted by the author,
Miss Lonsdale, towards the building. The classes of
cases received differ somewhat from those received in
other convalescent institutions ; they comprise :—

1. Persons to whom change of air is essential for
recovery, but who still require nursing, skilled dressing, and care ;

2. Cases still requiring surgical treatment ;

3. Persons needing bracing and preparation to
enable them to undergo operations ;

4. Children from five years old.

Incurable cases, children under five years of age,
cases attended with profuse and offensive discharge,
cases of pulmonary consumption beyond the first
stage of the disease, persons recovering from any
of the infectious fevers or contagious diseases, of
unsound mind, subject to epileptic fits, or of im-
moral character, are not eligible for admission.

The site is on a beautiful corner of Cannock Chase,
with a dry sandy and rocky soil, and in a fine bracing
air.

The building is of two stories and of a modest
unassuming appearance. On the ground floor are a
men's ward for six beds, a general dining-room,
matron's parlour with small dispensary attached, and

the kitchen offices. On the upper story are a similar
ward for women (six beds), a sitting-room for women,
matron's bedroom, two bedrooms for staff and linen
room. There is a hand lift close to the ward door.

The building has evidently been designed with the
most rigid attention to economy, and is certainly well
planned as regards both efficiency and economy of
administration. But it seems a pity that it should not
have been found possible to provide so essential an
adjunct as a bath-room. The cost of the building was
within a little of £2000, and of furnishing £250, and
the annual cost of maintenance is about £500.

Forres Leanchoil Hospital.—This cottage hospital
owes its existence mainly to the munificence of Sir
Donald A. Smith, K.C.M.G., of Montreal, Canada, a

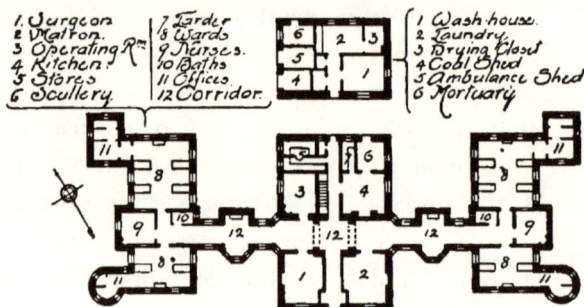

Ground Plan

FORRES LEANCHOIL HOSPITAL.

native of Forres. It is intended for the poor of Forres
and the surrounding district, the population of which is
about 10,000 ; of this about 5000 belong to the town.

The plan shows a central block, with two wings connected with the central block by corridors, and a small detached building at the rear of the central block. The central block is two stories in height, and contains on the ground floor the main entrance, with matron's and surgeon's rooms on either side; operation room, kitchen, scullery, larder, and stores behind; on the upper floor are bedrooms for the matron, nurses and servants. The connecting corridors between the central block and the wards are widened out in the centre, each with a bay window and a fireplace recess to form day-rooms for convalescent patients. The wings each contain two wards, one for four beds, the other for two beds, with a nurse's room and bath-room placed between the two. The larger wards are twenty-four feet long by twenty feet wide and thirteen feet high, giving a superficial area of 120 feet, and cubic space of 1560 feet per bed. The smaller wards are thirteen feet long by twenty feet wide, and give a superficial space of 130 feet, and a cubic space of 1690 feet per bed. The wards are lighted by windows in the side and end walls, fitted with double-hung sashes with fanlights over.

The wards are provided with ventilating stoves, especially designed for this building, supplied with fresh air from the outside, and the corridors are warmed by "Galton" stoves. For ventilating the wards, in addition to the windows, air shafts are fixed in the walls, fitted with adjustable hopper arrangements, at a height of ten and three-quarter feet from the floor. These hoppers, of which there is one to each bed, can be opened or closed at will. Exit ventilators for foul air are provided in the roof. The ward floors are laid with hard Canadian maple, in three-inch widths, wax-polished, and laid on deal counter flooring on wooden joists. The small detached building contains a wash-house and laundry, ambulance house and mortuary. The architect of the hospital was Mr. H. Saxon Snell, of London.

Watford Cottage Hospital.—Mr. Charles Ayres, of Watford, the architect of the Watford Cottage Hospital, containing nine beds, has produced a plan so compact and complete as to make it worthy of study. The building, which is most picturesque in outline, is of one story, and the beds are placed in two wards of four beds each, with an isolation ward for one bed. The lavatories are disconnected from the wards, and operation and bath rooms are added. The kitchen is cut off from the main building by a passage; and the whole plan, as will be seen, leaves little to be desired. The total cost of the buildings was, however, quite £1800, making the cost per bed £200. This is a large outlay, and in addition to this expenditure £550 had to be provided to defray the cost of the site, £221 for furniture, and £31 for instruments; making the total cost upwards of £2600, or nearly £300 per bed.

We are often asked the cost per bed of erecting cottage hospital buildings. Heretofore the pavilion plan has been so generally adopted that the initial

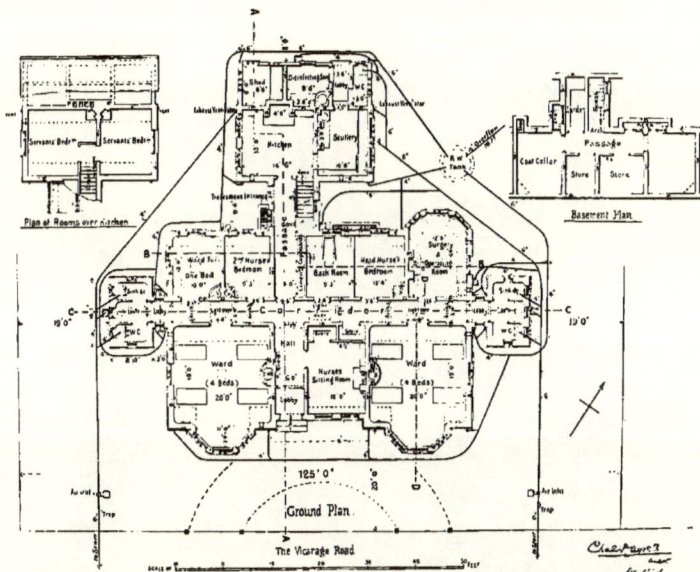

Plan of Rooms over Kitchen

Basement Plan

Ground Plan

125' 0"

The Vicarage Road

WATFORD DISTRICT COTTAGE HOSPITAL. *To face p. 264*

outlay on certain cottage hospitals has been relatively
enormous, and the working expenses afterwards are
often in consequence so great as to dishearten the
supporters of these institutions.   The cottage hospitals
at Stratford-on-Avon, at Darlington, and at Spalding
are specially notable instances of this type, and the
plans are worthy of study for this reason.   Our experience convinces us that in all cases where the
accommodation to be provided does not exceed ten
beds, on the whole it can best be provided in an
ordinary cottage, with slight and relatively inexpensive alterations.   We should urge this view to the
acceptance especially of wealthy people who may
desire to give a cottage hospital building to the
village or country district in which they take an
especial interest.   It is not wise or helpful to a
district to present a small community with a new
building beautiful in outline but expensive to maintain.   Where this has been attempted, with the best
intentions, by some wealthy landowner, it has not
unfrequently happened that the cottage hospital has
proved a burden too grievous to be borne, and thus
the original founders by their generosity do, in effect,
make the cottage hospital unpopular with the people
of all classes, and so cause its existence to be imperilled, if the institution is not ultimately closed as
too costly to be maintained by the limited population
to the needs of which it was intended to minister.

The plan which we here illustrate is quite a typical
one of what a cottage hospital ought to be.   It is

18

compact without being crowded, there is no wasted
space in passages, the whole of the accommodation
for patients and for nurses is on one floor, and the
kitchen offices are well isolated from the rest of the
building. The closets are properly separated from
the main building by cross-ventilated lobbies,—an
arrangement so often neglected in hospitals of this
type. There are two large wards, each for four beds,
and one small ward for one bed, making a total
accommodation of nine beds. Two bedrooms and a
sitting-room are provided for nurses ; and the servants'
bedrooms are placed in an upper story over the
kitchen. Each of the large wards has a spacious
bay window looking towards the south, which adds
much to the comfort and cheerfulness of the wards,
besides providing valuable space for convalescents. The
roof is projected over the recessed part of the front, between the two wards, to form a verandah. The treatment
of the exterior is simple and effective, and thoroughly
in keeping with the objects of the building.

The Memorial Hospital, Mirfield.—This hospital
(a plan of which appeared in *The Hospital*,
vol. xiv., p. 317, August 12th, 1893), built by
Mr. Charles Wheatley as a memorial of a deceased
sister, provides accommodation for sixteen patients,
and has cost, inclusive of site and furniture, £7300.
This is at the rate of £456 per bed ; but it should
be stated that a large, handsome room, now used
as a board-room, can, if necessary, be converted into
a ward, and that the necessity of making a road

to the gates of the hospital has added to the cost of
construction.   Mr. Wheatley, in addition to providing
and equipping the hospital, has promised an endowment of £100 a year towards the working expenses.
The hospital, a handsome building of freestone, facing
due south, stands on an eminence overlooking the
busy valley in which Mirfield lies.   Yet so well has
the site been chosen that green fields and trees,
ending in the woods which climb the hills on the
opposite side of the valley, almost completely blot
out the kilns, mills and long chimneys which here
abound ; and the surrounding atmosphere is clear
and bright.   The hospital is built upon a plateau
raised within its spacious grounds, and consists of
a central tapering, three-story building, from which
a wing, projecting slightly forwards, springs at each
end.   A verandah, covering an open tiled promenade
provided with seats, runs the length of the central
portion in the recess formed by the projection of
the end wings.   On the ground floor a corridor runs
the length of the building, and round this corridor
the hospital is built.   Entering at the front, we note
the carved oak doorway, the pretty coloured dado of
ornamental tiles in the lobby, and the oaken hat-stand,
before we reach the marble mosaic floor and white
tiled dado of the main corridor.   To the left of the
entrance is the doctor's room, to the right the
matron's ; then on each side an isolation ward of
one bed, with a cubic space of 2346 feet, and then
the general ward at each end.   On the other side of

the corridor are—kitchen : scullery, fitted with one of
Marsh's new stoves ; operation-room and surgery,
with marble mosaic floor, lantern light, as well as
window facing due north ; store-rooms, water-closets,
bath-rooms and lavatories. The general wards, one
male and one female, are alike in size and arrangement.
They are intended for seven patients each, with a
cubic space of 1426 feet per bed. The floors are
pitch pine, laid over concrete, with an air space
between. The walls are Parian cement, covered
with Aquol paint. Shorland's patent stoves are
fitted throughout the hospital, those in the general
wards being fixed in the centre of the room, with
descending flues. The windows are the ordinary sash
windows, worked by Meakin's sash openers. The two
general wards are well-lighted rooms, in which there
will be plenty of sunlight, and when the hospital is in
working order, when the floors have been polished,
and pictures, plants, with perhaps a canary, have been
added, they will be bright, cheerful rooms. The chairs,
tables, cupboards and bedside lockers, like the furniture throughout the hospital, are of oak ; and an
electric wire with detachable button reaching to the
patient's hands at the head of each bed enables each
patient to summon the nurse or matron at any
moment. A central chandelier, with ventilation
apertures in the ceiling, provides light. One feature
calls for comment. The water-closets and lavatory
of each ward are placed apart on the other side of
the main corridor. The ward projects into the main

corridor, narrowing it at the end, and is entered by
a door which looks not across but down the corridor.
Consequently, in going from the ward to the lavatories,
which are excellently arranged, it is necessary to pass
into the main corridor and round the projecting
corner of the ward, and it might perhaps startle
sensitive lady visitors or æsthetic young men to be
surprised by the appearance of a loosely clad patient
tripping from the ward to the lavatory; whilst we
are sure that a nurse bearing a bed pan would not
care to be confronted by visitors or others who might
be in the corridor. A secondary corridor with
through-ventilation shuts off the lavatories from the
main corridor, and if a door were made in the ward
wall opposite to it and opening across the main
corridor, not only would the disadvantage to which
we refer be overcome, but additional ventilation
would be secured for the ward. Four hot-water
coils heat the main corridor, and hot water circu-
lates through the metal towel-rails in the bath-
rooms, serving at once to warm the rooms and
dry damp towels. On the first floor are the board-
room, handsomely furnished in oak, bedrooms and
bath-room for the nurses; and on the second floor
is an attic for the servants. The sanitary arrangements appear to be perfect. Field's syphon flushing-
traps guard the heads of all outside drains,
which are all disconnected from service pipes. The
laundry and mortuary are placed in one small block of
building in the corner of the extensive back garden.

Falmouth Cottage Hospital.—So many very excellent cottage hospitals have been built of late
years, and so much publicity has been given to plans
of all kinds of hospitals, that we should have thought
it was an easy task to plan a hospital for ten beds

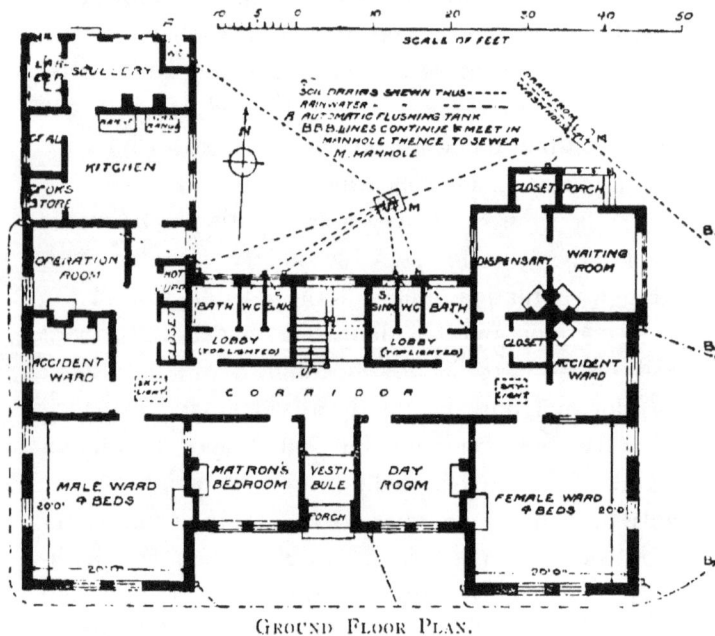

GROUND FLOOR PLAN.

without violating the first rules of hospital sanitation.
The present is, however, one of those cases which
show how difficult it appears to be to some architects to learn from what has gone before. Here is
an example of a one-story building with the corridor
so badly planned that top lighting has to be resorted

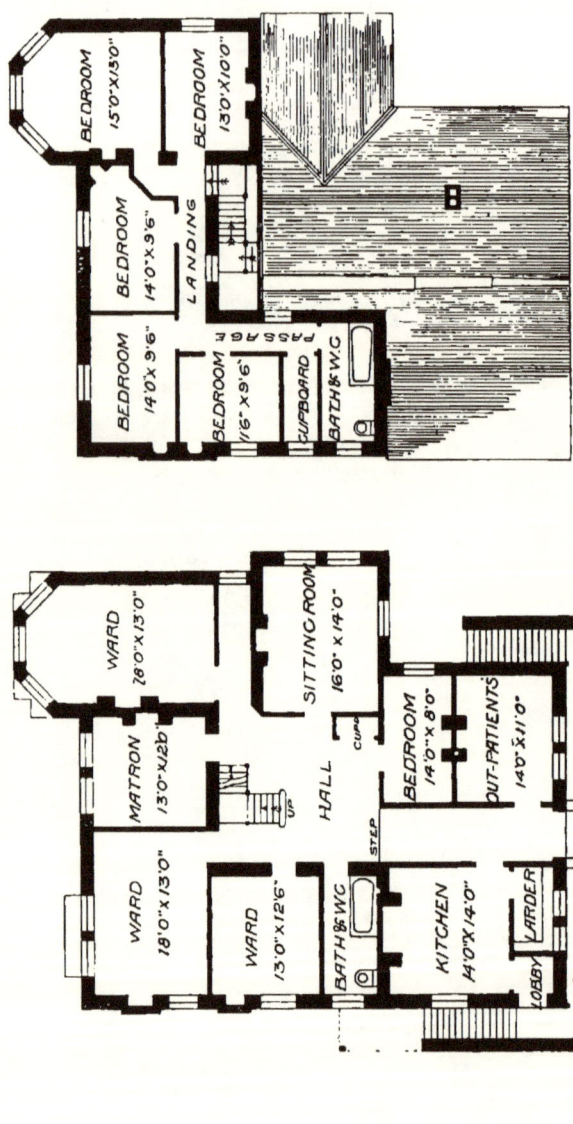

FIRST FLOOR.

GROUND FLOOR PLAN.   BRIXHAM COTTAGE HOSPITAL.

SCALE ... 10 5 0 10 20 30 40 50 60 70 OF FEET

First Floor labels:
BEDROOM 15'0"x13'0"
BEDROOM 13'0"x10'0"
BEDROOM 14'0"x9'6"
LANDING
BEDROOM 14'0"x9'6"
BEDROOM 11'6"x9'6"
PASSAGE
CUPBOARD
BATH & W.C.

Ground Floor labels:
WARD 28'0"x13'0"
MATRON 13'0"x12'6"
WARD 18'0"x13'0"
WARD 13'0"x12'6"
SITTING ROOM 16'0"x14'0"
HALL
CUP
BEDROOM 14'0"x8'0"
OUT-PATIENTS 14'0"x11'0"
BATH & WC
STEP
KITCHEN 14'0"x14'0"
LARDER
LOBBY
UP

to in order to get sufficient light ; with the water-
closets separated only by a top-lighted lobby from the
main corridor into which the wards open ; with bath-
rooms so small and approached in such a way, that
to carry a helpless patient in and bath him would
be an impossibility ; with the kitchen offices and the
operation room in closest proximity, and with the
out-patient room opening close on to two wards.
The operation room, which is very small, has its
window placed right up in one corner and close to
the fireplace, so that it would be impossible to put
the table opposite the window, and at the same time
have free space on both sides of it.

Brixham Cottage Hospital.—The plans we publish
of this building—designed by Mr. Tollit, of Totnes,
and erected at the expense of Miss Hogg as a
memorial to her sister and for the benefit of the
inhabitants of Brixham and its neighbourhood—
show the arrangements adopted on the ground and
first floors. The institution provides for nursing the
poor in their own homes, as well as such as come
beneath the roof of the hospital. Hence the upper
floor provides sleeping accommodation for nurses
largely in excess of what the ground floor seems
to demand.

The wards on the ground floor are intended respectively for men, women and children, the sitting-
room is for the use of the nurses, and the room
set apart for out-patients is intended also to be
used for operations in cases of necessity—a decidedly

unsatisfactory arrangement. A basement under a portion of the buildings provides additional larders, a scullery, and a rain-water tank. It is impossible to regard the disposition of the rooms as a successful example of hospital arrangement, even on so small a scale as this. The positions of doors, windows, and fireplaces in the wards do not seem to have been considered in relation to the places which the beds are to occupy. The arrangement, in one apartment, of baths and water-closets and the direct communication of this apartment and of the sanitary appliances with the main hall of the building are, unfortunately, retrograde, and should have been avoided. The dissociation of the scullery and kitchen, and the position of the latter with regard to the general entrance and the possible operating-room, are not judicious arrangements. In brief, the special requirements and risks of hospital work seem to have been insufficiently considered.

The Livingstone Cottage Hospital, Dartford.—This hospital is planned on a somewhat unusual model. It consists of two distinct buildings, one being the administrative block, the other containing the wards and their offices, the two parts being connected by a covered way. There is much to be said in favour of this arrangement : it keeps the wards absolutely separate from the administration offices, and permits of a freer circulation of air than would otherwise be the case.

The administration block is partly two stories in

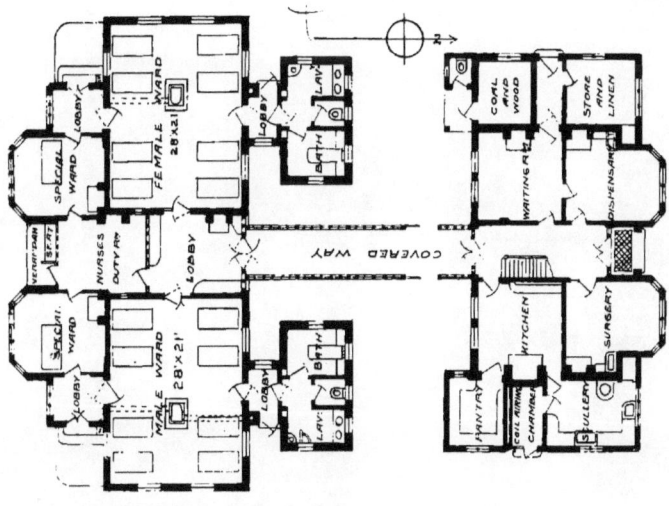

Scale of Feet

SPECIAL
WARD

FEMALE WARD
28'x21'

LOBBY

STAFF

NURSES
DUTY R'M

SPECIAL
WARD

LOBBY

MALE WARD
28'x21'

LOBBY

VERIFY DISH

LOBBY

LOBBY

BATH

LAV.

BATH

LAV.

COVERED WAY

COAL
AND
WOOD

WAITING R'M

STORE
AND
LINEN

DISPENSARY

KITCHEN

SURGERY

PANTRY

COAL ROOM

CHARNEL

SCULLERY

N

LIVINGSTONE COTTAGE HOSPITAL, DARTFORD.   [To face p. 172.

GROUND FLOOR PLAN.

height, the upper story being devoted to bedrooms
for the staff. On the ground floor the separate side
entrance to the waiting-room seems to suggest out-
patient work; the dispensary, if, as would appear by
there being communication between it and the waiting-room, it is used as consulting-room for out-
patients, is a convenient room for the purpose; otherwise it appears much too large for its work. The
surgery is, we presume, the operation-room, and is
provided with a top-light in addition to the windows.

The ward block contains two wards of eight beds,
each entered from a wide lobby, which, being provided
with a fireplace, may very well serve as a sitting-room
for convalescent patients. Between the wards is the
nurses' room, which serves the double purpose of
sitting-room and duty-room. On each side of the
nurses' room is a special ward for one bed. These
last each communicate on one side with the nurses'
room, and on the other with a lobby into which the
large ward also opens, and from which access to the
outside is gained. The planning of this block is not
altogether satisfactory. The lighting and ventilating
of the large wards are impeded by the position of the
special wards and lobbies; and the nurses' duty-room,
with its four doors, cannot be a very comfortable
apartment. The arrangement of the ward windows
is specially unfortunate, as the side where there is
least window space happens in each case to be that
facing south. In a small hospital like this, with

accommodation for only one nurse on duty at a time, the special wards might well have been dispensed with. Indeed, it would be a practical impossibility for one nurse to manage efficiently more than the sixteen beds in the two large wards.

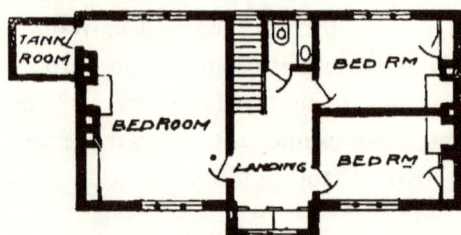

LIVINGSTONE COTTAGE HOSPITAL. FIRST FLOOR PLAN.

The hospital was designed by Mr. G. H. Tait, a local architect, and the first stone was laid on April 21st, 1894, by Mr. H. M. Stanley, the explorer.

Cottage Hospital, Wood Green.—The hospital stands on an open selected site, of seventy-one acres, fronting the Wood Green Road. It is approached by a carriage drive and is entered by a porch leading into a vestibule. On one side is the matron's sitting-room, with women's ward for four beds and a single room for one bed. The other side is similar for the men, including a day room. The two wings are connected by a wide and well-lighted corridor. In the centre is the kitchen block, with laundry in the rear and a mortuary.

An operating room is on the men's side, well-

lighted and fitted with every convenience. A
nurses' sitting-room and stove are provided near the

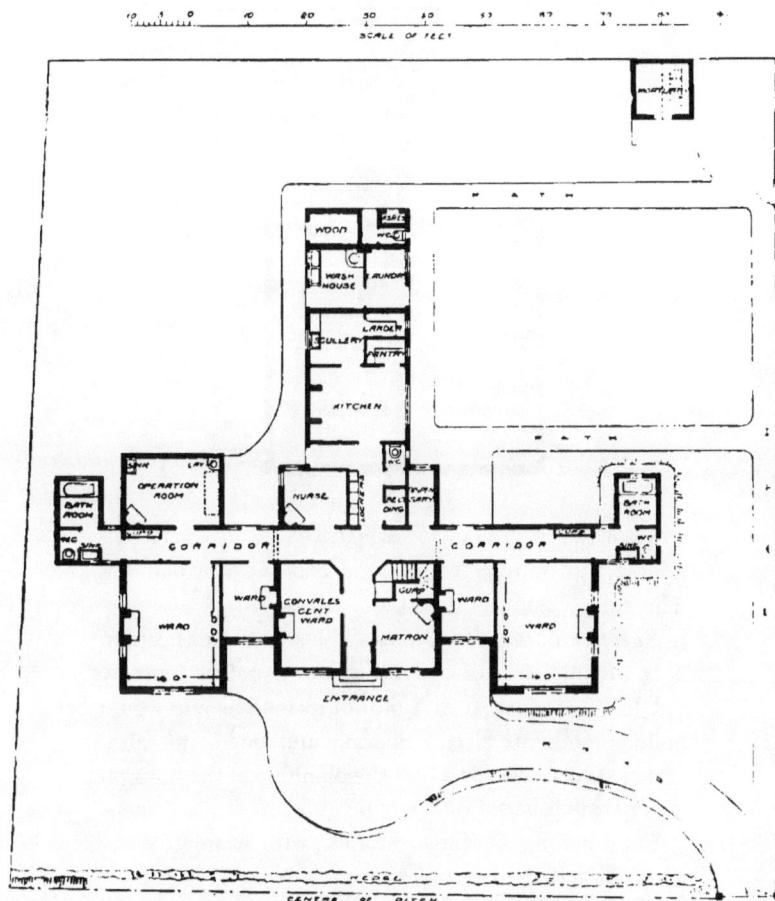

WOOD GREEN COTTAGE HOSPITAL.—GROUND PLAN.

centre. The water-closets and bath-rooms are each
placed in a separate and cross-ventilated block.

On the upper floor, over the centre, are the matron's and nurses' bedrooms. Extension on each side for

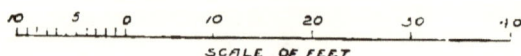

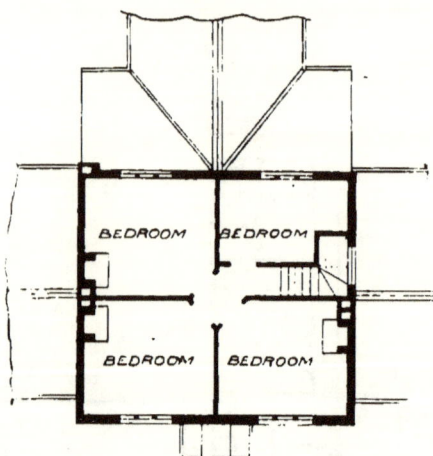

WOOD GREEN COTTAGE HOSPITAL.—FIRST FLOOR PLAN.

double the number of present beds is provided for. The total cost was £1800.

Surbiton Cottage Hospital.—The site of this hospital is an oblong strip of ground, about 450 feet long by 225 feet wide, just such a plot of ground as one would build a moderate sized villa on ; and one is not altogether surprised to find that the planning of the hospital is very much based on the villa type.

The building is of two stories, with a small basement for cellars and stores.

On the ground floor are a large day room for con-

SURBITON COTTAGE HOSPITAL.

SCALE OF 0 5 10 20 30 40 50 60 70 FEET

ARCHITECT FOR THE HOSPITAL
MR ERNEST GARRITT.

THE BUILDING CONTAINS
GROUND FLOOR
FIRST "

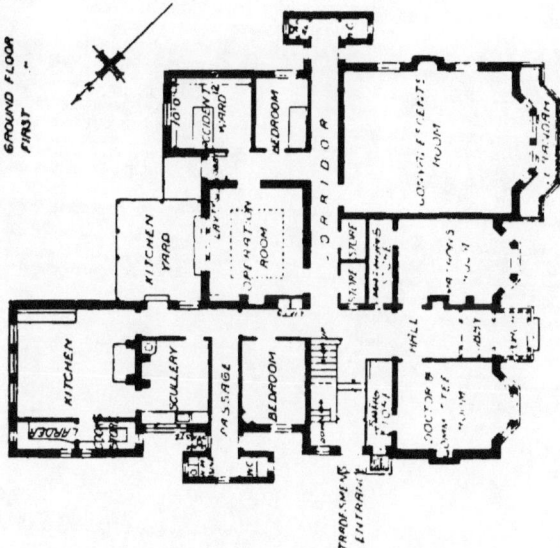

GROUND FLOOR PLAN

KITCHEN
LARDER
SCULLERY
PASSAGE
BEDROOM
KITCHEN YARD
LAVATORY
DISPENSARY
OPERATING ROOM
CLOAK? WARD
W.C.
BEDROOM
CORRIDOR
CONVALESCENTS ROOM
VERANDAH
HALL
DOCTOR &
SURGERY
WAITING
ROOM
TRADESMEN'S ENTRANCE

FIRST FLOOR PLAN

FEMALE WARD 18.0"
9.6
NURSES ROOM
PASSAGE
BATH
SPECIAL MALE WARD
BATH
CORRIDOR
SKYLIGHT
MALE WARD 25.0"
30.0

valescents, presumably for both sexes, an accident ward and an operation room, also a committee and doctor's room with drug store attached, the matron's

sitting - r o om, kitchen, offices, and two bed- rooms. On the upper floor over the convalescent room is a male ward for five beds, special ward over the accident ward, two female wards for three and two beds respectively, two bedrooms for nurses, a bath- room for each

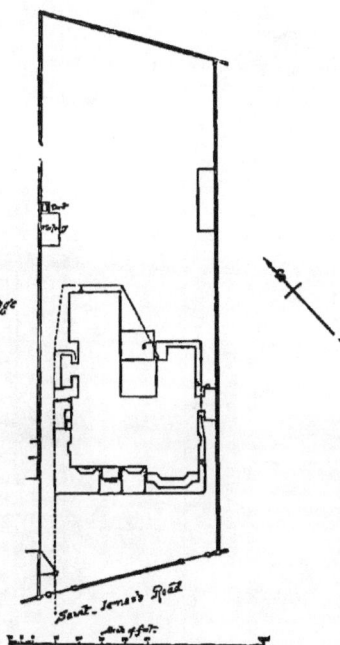

Surbiton Cottage Hospital.

Block Plan.

Saint Leonard Road.

sex and a small kitchen for night use.

The building was designed by Mr. Ernest Carritt and the cost was £3000.

Plans for the Erection of a Model Pavilion General, Fever, or Convalescent Hospital.—In conclusion, as a guide in some sense to those who wish to achieve efficiency of construction when deciding upon the plan of a new hospital, we have been urged to give a model embracing the chief features of im-

19

MODEL PAVILION GENERAL, FEVER OR CONVALESCENT HOSPITAL (WITH THIRTY TO FIFTY BEDS).

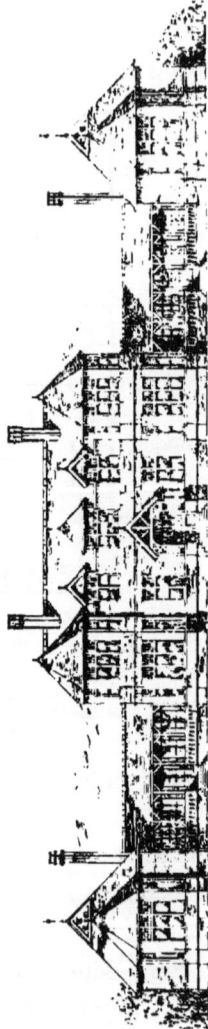

portance. In order to secure efficient drainage and sanitary arrangements, it will suffice to say here that the Grantham system, as described on pages 244 and 245, and also the instructions given in Chapter VIII., should be closely followed.

Our plan can be made available for a small or large institution, and it is capable of meeting the requirements of a fever, acute, or convalescent hospital. The authorities of St. John's Hospital, Northampton, have erected a convalescent institution of precisely similar design, with one exception, to the model given in the first edition. The exception is impor-

tant, for the architect carefully left out every
sanitary precaution, and whilst copying the main

PLAN OF MORTUARY.

GROUND FLOOR PLAN

MODEL PAVILION GENERAL, FEVER OR CONVALESCENT HOSPITAL
(WITH THIRTY TO FIFTY BEDS).

details of our plan he deliberately ignored all

precautions which were intended to secure the inmates from septic disease. Sewer gas was, in fact,
carefully laid on direct from the cesspool at each
available point, and the interior air became of the
worst possible description in consequence. Luckily,
the late Dr. Barr, of Northampton, had his attention drawn to these points before the building was
occupied, and so the patients were saved from what
would ultimately have produced disease if not death.
Imitation is, we know, the sincerest flattery ; but one
may be allowed to add that the appropriation even
of a hospital plan if unacknowledged should at any
rate be complete.

The accompanying model plan embraces all
the points of detail, and the whole of the principles of construction included in these pages. In
order to facilitate the description, each kind of
hospital will be separately detailed. In every
case the front central portion will have in the
basement stores, cellars, larder, etc., and on the upper
floor the bedrooms for the staff, nurses, and servants.

*Description of Small Cottage Hospital, with nine
or twelve beds.*—In this case the wings A and B will
be dispensed with, and the lavatories, etc., will communicate with the corridor by a cross-ventilated lobby.
Passing through the porch, which is fitted with seats,
a well-lighted hall is entered. On the right is placed
the matron's sitting-room, and on the left the
surgery or accident room, and beyond the matron's
room again is a well-arranged store-room of ample

proportions. Ascending the staircase we enter the sleeping apartments of the matron, nurses, and servants, which are spacious and admirably adapted for the purpose. The kitchen is seen from the windows at the back, and it is noticed that no room is placed over it, and that its roof is well ventilated and has an open skylight of considerable proportions. Thus no smell of cooking ever enters the hospital wards and passages. Returning to the entrance hall we enter room K, which can either be utilised as a ward for three beds or as a committee room. By another door beneath the staircase an area is reached which leads to a private water-closet for the staff. Communicating with the corridor are three wards, D, C, and E, each of which contains three beds. L is used as a dining hall or day-room where the convalescent patients are able to sit and to have their meals. F is the kitchen, and H, which communicates directly with the entrance hall, is a well-lighted operation theatre, replete with all the necessary appliances and fittings. It will be seen that this plan admits of an arrangement which will give twelve beds for the use of patients, by sacrificing the committee room, which can well be dispensed with, as either the matron's room or the surgery could be made available for such a purpose. In practice this plan has much to recommend it. The cost of building such a hospital is estimated at £3000 ; furniture will cost £200, making a total expenditure of £3200.

Situated in its own grounds, upon a well-chosen

site, such a hospital building leaves little to be desired.

General Hospital, with thirty or forty-eight beds.—The foregoing description of the central block would apply in this case also, and eighteen additional beds will be obtained by adding at each end of the corridor the two one-storied pavilions A and B, or thirty-six extra beds will be available if these pavilions are two-storied. The cost of such a hospital is estimated at £4000 for building, £470 for furnishing ; total, £4470 for thirty beds ; or £4600 for building, £650 for furnishing ; total, £5250 for forty-eight beds.

A Convalescent Institution, with fifty beds.—This plan has been carried out at Weston Favell, Northamptonshire, with perfect success, and a very complete convalescent institution with thirty beds may there be seen. It will well repay a visit, and can easily be reached from Northampton, as it is within an easy drive of that town. To make our plan available for convalescents it will require slight modification in one or two particulars. Thus A and B with floor above will easily accommodate thirty-six patients, C and D can be used for sleeping rooms with four beds in each, and K and L will easily accommodate three patients each. A splendid dining-hall and general sitting-room is obtained by throwing E and H into one apartment, and its shape well adapts it for such a purpose. G would again be used as the medical officer's apartment, and the other arrangements of the central block would remain as in the first case.

The cost of erecting such a building is estimated at
£6000, and the furniture will entail an additional
expenditure of £600 ; total, £6600.

In all these plans the position of the lavatories and
water-closets will depend upon the size of the building and the purposes to which it is devoted. Thus,
although these offices are shown in the plan, both as
connected with the pavilions A and B and also with
the corridors, where the pavilions are erected, it
would be unnecessary to have the corridor lavatories.

It will be noticed that the kitchen is centrally situated, and adjacent to it is the bath-room (not marked)
so as to be the more economically heated. One bath
would be sufficient for a hospital of twelve beds, but if
on a larger scale, a bath-room should be built out with
the lavatories, etc., at the end of the pavilions A and
B. A small ward with three beds is conveniently
placed attached to the operation theatre, the aspect
and general principles applying to the larger wards
being strictly preserved.

The corridors, by an arrangement for completely
throwing open the windows, would become in summer
practically nothing more than covered ways, and
verandahs are provided, which, if thought desirable,
might be carried across the front of the centre building.

The wards in the pavilions A and B, each containing nine beds, are placed at the end of the corridors.
A bay window is thrown out at one end of each,
which is not only a convenient arrangement, but also
gives a remarkably cheerful appearance to the wards.

It will be noticed that the water-closets, lavatory, sink, etc., are invariably cut off from the wards and corridors by a lobby or separation with cross ventilation. We cannot too strongly impress upon architects and others who are responsible for hospital construction that this plan must invariably be adopted, because, however perfect may be the drainage and apparatus, experience proves this lobby to be a *sine quâ non* in hospital construction. A plan is given of a mortuary with *post-mortem* and inquest rooms, which, with the wash-house, laundry, etc., would be detached buildings placed according to the requirements of the site. Such a mortuary as that given in the plan would cost from £200 to £300.

Although these plans give an appearance of a somewhat extensive building, it will be found on comparing them with the other plans given in this book that the cost of erection is in no case excessive. Abundant roof space is most desirable from a hygienic point of view, and it becomes an important consideration when the water supply is either uncertain or limited. This is more often than not the case in country districts, and so we have carefully considered it in preparing these plans. In every case the roof space is ample to provide against any shortness in the water supply, if proper tanks are provided and all the rain-water is carefully stored.

Fever Hospital.—The central block with twelve beds, already described, would answer well as the nucleus for a larger number of cases, if tents or sheds were erected, as occasion might require, on the plan described in Chapter VIII. Such accommodation does not cost a large sum, the tents or sheds are easily erected and removed, and by their aid an epidemic can be met with promptness and success.

Permanent Fever Hospital, with twenty-four or forty-five beds.—In this case A would be used as the male fever ward with nine beds, B as the female ward with nine beds, C and D as the male and female convalescent rooms, E, H, K, and L as isolation wards, and G as the medical officer's room. If A and B are two-storied pavilions, eighteen additional beds would be available, and the central block would still be devoted to the purposes previously detailed. This plan has many advantages, because A and B might easily be utilised for different kinds of fever, a separate entrance being obtained to each through the corridor at the back of the hospital. The cost of erection, owing to the extra cubic space required for these cases, would be one-third more at least, say £5000 with one-storied pavilions, or £6500 if they were of two stories, and the furniture would cost about £600; making a total expenditure of £5600 in the first case, or of £7100 in the second.

Having now had the advantage of seeing these model pavilion hospitals in working use, the author feels justified in recommending them as well adapted to the purposes for which they are intended. Any one who doubts this can easily judge for himself by paying a visit to St. John's Hospital, Weston Favell, Northampton.

Isolation Hospitals—Local Government Board's Model Plans.—In January, 1895, Dr. Thorne Thorne, the medical officer of the Local Government Board, issued a memorandum on the provision of isolation hospital accommodation by local authorities, to which three model plans were attached. All these plans provide for movable baths and earth commodes for the wards. It will be seen that accommodation for the nurses is provided for in the caretaker's cottage, or in an upper story of the ward block. In every plan the specification provides for a close fence, 6 feet 6 inches high, all round the site ; and that the distance between each of the buildings is at least 40 feet, the distance to the boundary being also in every case 40 feet. In all the plans provision is made for 2000 cubic feet of air space, 144 square feet of floor space, and 12 linear feet of wall space per bed as a minimum. The Board hold that water-closets are preferable to earth-closets where efficient sewers are available; but in plan A (see opposite) earth-closets are indicated as the means of excrement disposal. In the selection of a site for an infectious hospital, the following considerations are of primary importance—the wholesomeness of the site, the character of the approaches, and the facilities for water supply and for slop and refuse removal. It is insisted that the 40 feet of interval between the buildings and the boundaries should not afterwards be encroached on by any temporary building, or for

PLAN A.

LOCAL GOVERNMENT BOARD
MODEL PLAN FOR FOUR BEDS

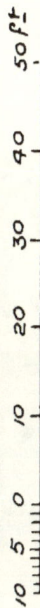

a.a. THIS DISTANCE SHOULD BE 40'0"

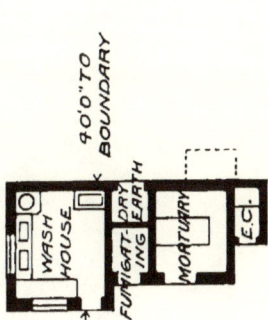

GROUND FLOOR PLAN

Labels within the plan:

WASH HOUSE

FUMIGATING · DRY EARTH

MORTUARY

E.C.

40'0" TO BOUNDARY

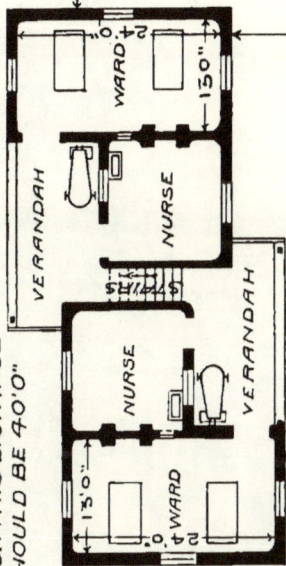

VERANDAH

WARD 2 4'0"

15'0"

NURSE

STAIRS

NURSE

VERANDAH

15'0"

WARD 2 4'0"

40'0" TO BOUNDARY

90'0"

a

a

ADMINISTRATION COTTAGE

A. STAIRS TO NURSES' BED ROOMS

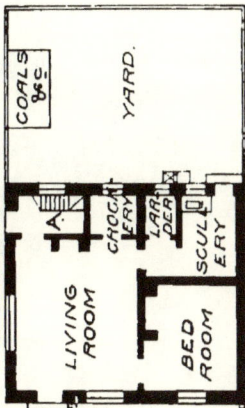

LIVING ROOM

CROCKERY

LARDER

SCULLERY

BED ROOM

COALS &c.

YARD.

extensions in connection with the hospital; and that where temporary buildings are required, the same principles should be kept in view.

It cannot fail to be helpful to first deal with the plans as a whole, and then to refer to them more in detail. Attention must first be directed to the fact that Plan C is only a ward block for one disease in the two sexes; so that it would not alone suffice for patients' accommodation in any isolation hospital, since it fails to supply means of isolating a second disease, or a doubtful case, or a non-infectious case sent in in error wrongly diagnosed, or a case in which, *e.g.*, scarlatina is concurrent with measles, or is complicated with diphtheritic symptoms, all of which contingencies occur not infrequently. Owing to the foregoing conditions, the block shown on Plan B, or some modification of it, must be adopted as indispensable for every isolation hospital. Those who have to erect an infectious hospital must bear in mind that the accommodation provided in the administration block must be adequate for all nurses and servants who may be needful in the event of the hospital being in full use with two or more infectious diseases. The employment of temporary non-resident helpers involves too grave a risk of their spreading infection to their homes to be permissible. Another point, often overlooked, is, that complete laundry arrangements are essential to the safety of the community; and all such arrangements must be more than ample, or they are sure to break down in times of pressure.

The first model, Plan A. is that of a small building providing accommodation for two patients of each sex. Dr. Thorne Thorne's memorandum [1] gives a pleasing elevation of the building, which has an upper story, containing bedrooms and offices for the accommodation of nurses. It is calculated that the cost of erecting such a building as that shown in Plan A would be about £1500.

The second plan, or Plan B, provides a building with accommodation for eight patients, and makes provision for the separation of the sexes, and also of one infectious disease from another. A convenient disposal of buildings upon site is also indicated on this plan. The ward block on Plan B costs about £1300 ; or the whole of the plans complete, including the administrative block, laundry, etc., would entail an expenditure of about £3000.

The third plan, Plan C, is a small pavilion hospital, with accommodation for six male and six female patients suffering from one kind of infectious disease. The cost of the building would be about £2000.

Any one who is interested in this question should certainly purchase a copy of Dr. Thorne Thorne's memorandum, and study it carefully.

Hospitals for Small-pox.—Dr. Thorne Thorne points out : " Ever since the issue in 1882 of the report of the Royal Commission on Small-pox and Fever Hospitals

---

[1] *On the Provision of Isolation Accommodation by Local Authorities.* London, 1895: Eyre & Spottiswoode, East Harding Street. E.C., and of any bookseller. Price 1d.

PLAN B.

LOCAL GOVERNMENT BOARD
MODEL PLAN FOR EIGHT BEDS

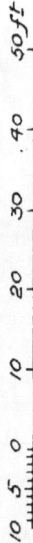

REFERENCE FOR BLOCK
PLAN.

A. ENTRANCE
B. ADMINISTRATION COTTAGE
C. WARD BLOCK
D. MORTUARY LAUNDRY &c.

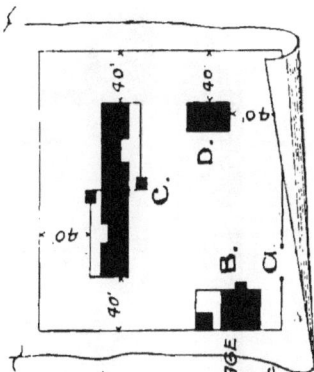

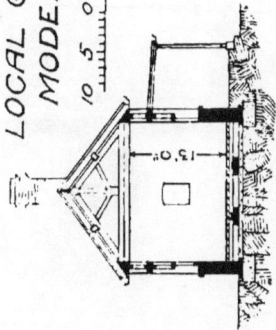

SECTION ON LINE A.B.

BLOCK PLAN

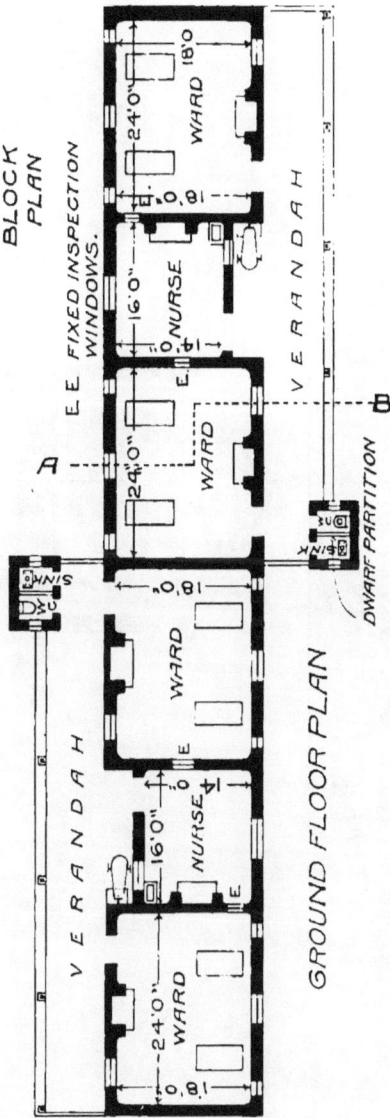

GROUND FLOOR PLAN

E. E. FIXED INSPECTION WINDOWS.

VERANDAH

WARD 18'0

NURSE

WARD

WARD

NURSE

WARD

SINK

DWARF PARTITION

PLAN C.

LOCAL GOVERNMENT BOARD
MODEL PLAN FOR TWELVE BEDS

10 5 0   10   20   30   40   50 ft

SECTION on line A.B.

13' 0"

GROUND FLOOR PLAN

WARD

STOVE

26' 0"

36' 0"

NURSES ROOM

LINEN

STORE

BATH

W.C.

WARD

STOVE

26' 0"

36' 0"

A

B

S.SINK

W.C.

W.C.

S.SINK

great difficulty has arisen in the selection of sites for
the reception of patients suffering from small-pox.
Small-pox hospitals have again and again served to
disseminate that disease to neighbouring communities,
and this, to use the words of the Royal Commission,
'in spite of precautions almost in excess of any that
could have been anticipated.'" A local authority
should not erect a small-pox hospital—"(1) On any site
where it would have within a quarter of a mile of it as
a centre either a hospital, whether for infectious diseases or not, or a workhouse, or any similar establishment, or a population of 150 to 200 persons ; (2) On
any site where it would have within half a mile of it
as a centre a population of 500 to 600 persons whether
in one or more institutions or in dwelling-houses."
Even these precautions may not suffice in certain
circumstances. Hence the Local Government Board
for some time past have refused to sanction loans for
isolation hospitals unless the local authority has first
given an undertaking that cases of small-pox will
not be received or treated on the hospital site at
the same time that cases of any other infections
are being isolated there.

The suggestion that small-pox hospitals may be so
constructed as not to be dangerous to neighbouring
habitations, by a system of passing through a furnace
all outgoing air from infected wards and places,
though reduced to practice, Dr. Thorne Thorne
states "cannot be regarded as having sucessfully
attained the end in view."

The Construction of Small Cottage Hospitals.—In
our last edition we drew attention to several new cottage hospitals designed upon the pavilion principle on
the lines already laid down in this chapter.   We unfortunately omitted to caution the inexperienced that the
pavilion system is too costly to be extended to hospitals
having less than 50 beds, and, as a matter of fact, several
of the more recent cottage hospitals demonstrate this
fact beyond all question.   The importance of economy
in cottage hospital construction is generally admitted ;
and at the request of correspondents in England and
America, of architects and hospital authorities, we
are induced to suggest some alternative plans based
upon an exceptional experience, combined with a
careful consideration of the faults apparent in several
of the more ambitious cottage hospitals which have
been erected during recent years.

The ends to be kept in view in building a cottage
hospital are : (1) economy in administration ; (2)
convenience of intercommunication between every
portion of the building, so that the nurses, matron
and her assistants may be saved as much labour as
possible ; and (3) in cases where the population is
increasing, facilities for extension, should occasion
arise for increasing the hospital accommodation.   All
these points have been considered ; and we are of
opinion that the plan which seems preferable to the
pavilion for small cottage hospitals is one designed
upon the straight model.   Such a plan enables the
architect to arrange all the wards upon the ground

floor, and to provide perfect ventilation. The matron's and nurses' rooms, and the bath and operating rooms can also be placed on the same floor; whilst the kitchen, with sleeping accommodation for the staff, can be located upstairs. A lift communicating with the kitchen and ground floor should be provided. The complete isolation of the kitchen, when necessary, and the avoidance of many evils, which might otherwise arise, may be secured by the erection of a special staircase connecting the outside and back portion of the premises with the kitchen, by which access could always be obtained to the latter without entering the hospital. Such a plan affords the widest scope for future extensions, should they be necessary at any time.

Some people have a strong prejudice in favour of keeping the sexes apart, by placing the female wards on the first floor. They also prefer a two-story building, on the ground that it is less pretentious, and more home-like and comfortable, than any other which can be devised. We have endeavoured to meet the views of such persons in Plan No. 3, and have been induced to give such a plan with the object of showing that, although the building occupies less ground, and would, therefore, at first sight be thought to entail a less expenditure, in practice it will be found that this form of building is as costly as the other.

We find that whoever desires to secure a cottage hospital for a particular locality is impressed with the feeling—and we think rightly so impressed—that the elevation of the building, however plain, should at

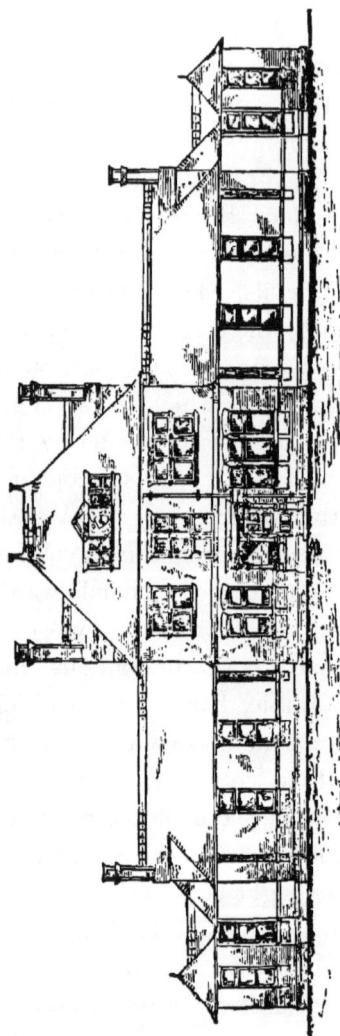

Burdett's Straight Model. Plan 2.—Elevation.

least be attractive and striking. We therefore commence our series of plans by giving the elevation of ground Plan No. 2, in order to show that the straight model is not unsightly, but the contrary, as will be seen from the sketch in margin.

Plan No. 1. Suitable for an Infectious Hospital.—The chief feature in the first design is the entire separation of the administrative portion of the building from the ward block, the two being connected by a covered way. The advantage of this plan is that the ward or wards and the administrative block can be placed in any position on

the site, and in relation to each other.   It also most
effectually separates the quarters of the rest of the
staff from the ward
atmosphere.   For
these reasons the
plan is commended
to the consideration
of local sanitary au-
thorities and others
who have to main-
tain infectious hos-
pitals where the
population is com-
paratively   small.
There is no reason
why new blocks on   Burdett's Straight Model. —Plan i.
this plan should not be indefinitely extended to meet
the requirements of special localities, and at a comparatively small cost.

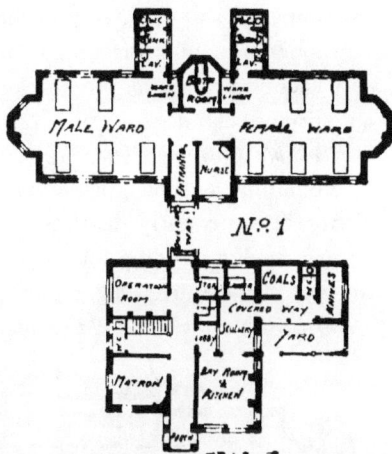

Administration Block.—Here will be found the day-
room and kitchen directly communicating with a
covered way, and adjacent to it a scullery, larder,
coals, etc.   On the left of the entrance are the matron's
sitting-room, and the staircase, with the operation-room
beyond, whilst opposite to the latter are the rooms for
linen and stores.   Upstairs there are provided five
bedrooms for the matron and staff.

Ward Block. —This is self-contained, and is arranged
to provide accommodation for five male and five
female patients.   Each ward has a disconnected

lavatory, sink, and water-closet, with a separate linen
closet, and other conveniences. There are also a nurses'
room and a bath-room, both of which are placed in
the centre, so as to be easily available for both wards.
An abundance of cubic and floor space is provided,
and each ward would present an unusually cheerful
and airy appearance.

Probable Cost of Plan No. 1, Exclusive of Site.—
The building, as shown in the plan, cubes up at 6d. to
£2010, but in country districts it ought to be possible
to erect such a building for about £1500, exclusive of
the cost of furniture.

Burdett's Straight Model Cottage Hospital, Plan No. 2.

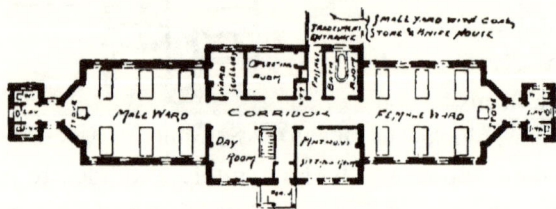

GROUND FLOOR.

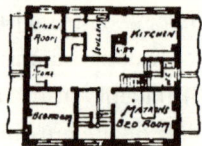

FIRST FLOOR.
BURDETT'S STRAIGHT MODEL. COTTAGE HOSPITAL FOR
TWELVE BEDS.

— In this plan the kitchen is placed on the top floor, and,
in order to obviate the necessity for tradesmen's boys
and men entering the hospital, a lift is placed in a central

position, which serves equally well for carrying up
stores to the kitchen and for sending down meals to
the wards.   This lift is accessible on the ground floor
both to the ward corridors and to the tradesmen's
entrance.   It will be observed, however, that the plan
is so designed as to secure a complete separation between the tradesmen's lobby and the communication
with the wards.   A speaking tube on the ward side of
the lift would ensure ready communication between
the kitchen and ground floor.

Administration.—This is placed in the centre of the
building, and consists, on the first floor, with the
exception of the matron's bedroom, of a kitchen,
with scullery, stores, and laundry, the former of which
is connected with the wards and tradesmen's entrance
by a lift, as before described.   A linen room, the
nurses' and matron's bedrooms, and store closets, are
provided, whilst sleeping accommodation for the
servants and two nurses, if necessary, will be found on
the third story.

Ward Block.—This provides accommodation for
twelve beds in two wards of six beds each.   These
wards provide ample accommodation, and are entirely
disconnected from the offices by a cross-ventilated
lobby.   In cases where it is probable that an extension of the building may be found necessary, the
lavatories, etc., might be placed at the side, behind,
and at one end of the wards, as shown in Plan 3.
This might entail the sacrifice of one bed, but would
enable the beds to be doubled at the smallest possible

outlay. In those places where more accommodation
is not likely to be wanted, we should strongly advise
the adoption of the arrangement shown on the plan.
A spacious day-room is placed at the left of the
entrance, which is thrown open to the corridor, on the
model of the Halstead Cottage Hospital, where this
arrangement has proved in practice a success. A
ward scullery, an operating-room, and a bath-room
are also provided. Glass doors are placed on the
right of the entrance lobby, shutting off the corridor
communicating with the female ward from the day-
room and mess corridor. This plan, cubed out at 6d.,
would cost £1778, or less than £150 a bed, although
it is replete with every modern appliance, and would in
practice be found to leave little, if anything, to be desired.

Burdett's Villa Hospital, Plan No. 3.—In this plan the
kitchen offices are on the ground floor, a point of
some importance being a passage which separates
them from the rest of the hospital. This passage is
ventilated at each end, and would certainly tend to
prevent any smell of cooking from penetrating to the
front part of the building.

Ground Floor.—This provides accommodation for
five beds in the male ward, situated on the left of the
entrance, the scullery and offices being disconnected
from the wards and placed outside the building. The
matron's room is situated on the right of the entrance,
next to it being the staircase; then the bath-room,
having the linen closet in front, and an operating-
room with large window at the back. Beyond, again,

but disconnected from the patients' portion by the passage already referred to, are the kitchen or day-room, with the usual offices.

First Floor Plan. —This contains a female ward for five beds, similar to the male ward, with matron's bedroom, ward scullery, four other bedrooms, one of which might be used for isolating a special case, a linen room, etc. We do not like this plan half so well as No. 2, and we have only given it for the reasons already stated. It cubes up at 6d. to £1988; but in some parts of the country such a building might be carried out at 4½d. to 5d., which would bring the cost down to £1500, exclusive of furniture.

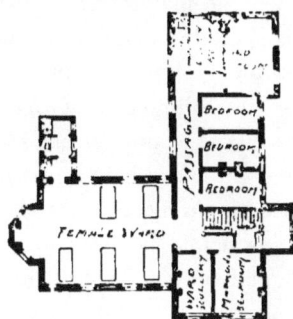

FIRST FLOOR

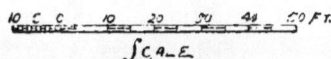

SCALE

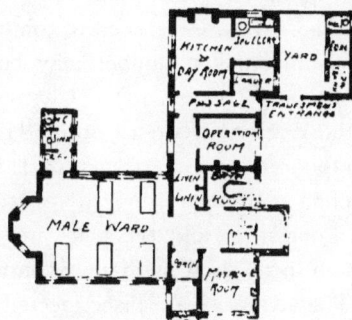

GROUND FLOOR

BURDETT'S VILLA HOSPITAL PLAN.

General Remarks.—Such are the three plans upon

which cottage hospitals, complete in all particulars, may
be erected and maintained at the smallest cost.   Plan
No. 1 is specially adapted for infectious cases.   No. 2
is undoubtedly the best for small cottage hospitals,
because the original cost, as well as the cost of
administration, would here be reduced to a minimum.
We are often asked how to estimate the cost per bed
of erecting buildings of this character.   Heretofore the
pavilion principle has been so generally adopted that
the original outlay on cottage hospitals has proved
enormous, and the working expenses afterwards are
often in consequence so great as to dishearten the
supporters of these institutions.   The hospitals at
Stratford-on-Avon, at Darlington, and at Spalding are
especially notable instances of this ; and the plans are
worthy of study for this reason.

It will be observed that the smallest of the three
hospitals for which plans are here given contains ten
beds.   We are disposed to think that accommodation
for any smaller number may best be provided in an
ordinary cottage.   Mr. Charles Ayres, architect for
the Watford District Cottage Hospital, nine beds, has
produced a plan (see page 264) so compact and complete as to make it worthy of study.   The building is
of one story, and the beds are placed in two wards of
four beds each, with an isolation ward for one bed.
The lavatories are disconnected from the wards, and
operation and bath-rooms are added.   The kitchen is
cut off from the main building by a disconnecting
passage, and the whole plan leaves little to be desired.

CHAPTER XI.

COTTAGE HOSPITAL APPLIANCES AND FITTINGS.

Surgical instruments—Dispensary requisites—Hair mattresses
—Iron bedsteads—Linen and blankets—Counterpanes—
Pictures in wards—Walls of wards—Lint and medical
sundries—Patients' necessaries—Lockers—Brackets—
Headings for beds—Ward furniture—Movable closets and
baths—Hot-water plates—Screens—Bed rests—Arm slings
—Surgical hammocks—Easy chairs—Foot rests—Filters—
Book shelves—Ambulances—Miscellaneous articles—Bags
and pads—Hot-water tins—Feeding cups—Bed pans.

OUR description of cottage hospitals and their management would scarcely be complete unless some hints were given as to the kind of appliances and fittings most suitable for cottage hospital purposes, and the best means of obtaining them. It is, of course, necessary to go much into detail in referring to this branch of our subject. The following list of requisites for the wards may be found useful.

Instruments.—A good collection of surgical instruments is one of the chief requisites for a cottage hospital. The instruments should be kept for the use of the medical men in the neighbourhood, with certain restrictions to ensure their safety and cleanliness. As regards the expense of starting such a collection, we would advise that, at first, only the necessary instruments be purchased, either by means of a separate subscription, or, if possible, by funds raised from a

(299)

generous donor. A fairly complete set of instruments being once obtained, the best plan of increasing it, and keeping it in good repair, will be by levying a certain annual subscription on each medical man who participates in its benefits. We cannot advise that water beds, bed rests, etc., be included in the list of instruments, and that persons be allowed the use of them on paying a certain subscription. Such a plan will entail a great deal of extra trouble and annoyance to the working staff, and is quite foreign to the objects for which a cottage hospital is established.

Appended is a list of instruments which will be found useful, with their probable cost.

It will be noticed that it is far from complete. The aim in drawing it up has been to include instruments which will certainly be required, and also such as, although only occasionally wanted, are likely to be required in a hurry whenever the occasion arises. On the other hand, many instruments, the necessity for which is not likely to arise as an emergency, are omitted, because, as for example in the case of lithotrity or lithotomy instruments, there will always be time to either purchase or hire them. In regard to some of the less common operations, hiring will not infrequently be found the best plan, not only because fashion changes, but because individual operators may have their preferences ; and it is unfair that in important operations they should have to use what they may consider inappropriate tools. Fortunately, the instruments of emergency are mostly comparatively simple.

List of Instruments.

Case of Amputating Instruments, say	£5	0	0
Case of Knives, etc. :—			
4 Scalpels			
2 Bistouries ...			
1 Hernia Knife ...			
2 Tenotomy Knives	3	0	0
1 Aneurism Needle			
1 Nævus Needle ...			
1 Tracheotomy Hook ...			
Case of Trephining Instruments :—			
Trephines, Hey's Saw, Horsley's Elevator, Skull Forceps, Scalpel	2	10	0
Esmarch's Elastic Tourniquet	0	7	0
Cutting Bone Forceps	0	12	6
2 Necrosis Forceps	0	14	0
Lion Bone Forceps	0	7	6
Chisel	0	4	6
Mallet ...	0	4	6
2 Scoops	0	14	0
Gouge	0	5	0
6 Pressure Forceps	1	10	0
Sinus Forceps ...	0	3	6
Vulsellum Forceps ...	0	7	6
Polypus Forceps ...	0	6	6
Hernia Director... ...	0	4	0
Ordinary Director	0	3	0
2 Probes	0	2	0
2 Scissors	0	7	0
1 Spring Dressing Forceps	0	2	0
2 Metal Retractors	0	6	0
Esmarch's Irrigator	0	7	6
4 Glass Drainage Tubes (for Abdomen)	0	4	0
3 Tracheotomy Tubes	2	5	0
4 Sponge Holders	0	8	0
1 Lithotomy Sound	0	7	0
½ dozen Silver Catheters, with 1 Prostatic in case...	1	12	6

Carry Forward £22 14 6

Brought Forward	£22	14	6
1 dozen French Elastic Catheters, à boule	1	0	0
3 Elastic Condée Catheters	0	7	0
Hydrocele Trocar, with Syringe in case ...	0	13	0
Throat Probang...	0	2	6
Billocq's Epistaxis Canula	0	5	0
Ear Syringe and 3 Canulæ ...	1	0	0
Aspirator...	1	10	0
Clinical Thermometer	0	3	6
Hypodermic Syringe	0	3	6
Enema Syringe ...	0	3	6
Uterine Sound	0	5	0
Tooth Instruments, say	4	4	0

Needles, Harelip Pins, Silver Wire, Silk⎫
    Ligatures, Catgut Ligatures in bottles,⎪
    Rubber Drainage Tubes in bottles.⎬ say ...   2   0   0
    These will require constant renewal⎭

Splints :—

3 Neville's Back Splints for leg	1	7	0
2 Lawrence's Bed-cradles	0	17	0
3 Angular Splints for the elbow, with reversible metal joints	0	12	0

Paste Board, Plaster of Paris, and⎫
    Plaster of Paris Bandage ...⎬ say ...   0 10   0

All Straight Wooden Splints are best made as required by the local joiner, and when so made cost very little indeed.

Dressings :—

Lint, Tow, Strapping, Bandages, Absorbent⎫
    Wool, Cyanide Gauze, Iodoform Gauze,⎪
    Oil Silk, Jaconet, Sponges. All these⎬ 5   0   0
    will require constant replenishing ...⎭

Ether Inhaler	2	2	0
Steriliser [1] for Instruments and Dressings	5	10	0
	£50	9	6

[1] Unless surgeons are prepared to carry out in their entirety the details of anti- as distinguished from aseptic surgery, a

Dispensary Requisites.—These are—shelves, drawers, bottles, ointment pots, desk and writing conveniences, label tray, scales, cupboards for splints, etc.   It is advisable that a separate kind of bottle should always be used for lotions, liniments, and other outward applications.   For this purpose, the blue fluted bottles are extremely useful, as they can be felt in the dark, and with such bottles a mistake is almost impossible.   There are many other varieties and patterns, accounts of which will be found in *The Hospital* newspaper.   All poisonous drugs must be kept in a separate locked cupboard, and any medicines containing poison should be dispensed in a ground glass bottle, as when touched it will ensure caution.   There should also be conveniences for testing urine—test-tubes, spirit-lamp, urinometer, reagents, test-papers, etc., and a white earthenware sink should be provided.

Hair Mattresses.—These are much more economical in the long run than those made of any other material, and for practical purposes they are to be strongly

steriliser is absolutely necessary; and even with the most complete chemical antisepsis it is a very great advantage.   It should be borne in mind, however, by the ingenious surgeon that a Warren's cooking pot and some empty biscuit tins, costing about 15s. altogether, can be made to serve much the same purpose.   To this should be added at least £6 10s. for a small glass and metal instrument cupboard, and something should also be allowed for eye and throat instruments and for things required for certain abdominal operations, all of which depend on the special aptitude and inclination of the surgeons of the institution.

recommended above all others. They should be at
least 4½ inches thick. We strongly recommend that
they should be made by the nurse, and that the hair
should be purchased separately ; as in this way, when
a bad case is admitted, we may be more sure that the
mattress is taken to pieces, and the hair properly
washed, disinfected, and re-picked before the mattress is
again filled. Hair of good quality may be obtained at
a considerable reduction from H. M. Dockyard, Woolwich, through any broker. The hair may be purchased in this way at from eleven to thirteen pence
per lb., of the best quality, and of absolute purity.
The Government sell all the materials which have
been in use during a cruise, whenever a ship is paid
off; and thus it is that good hair, nearly new, thoroughly disinfected and purified, may be obtained for
a nominal outlay. Should it be thought too much
for the nurse to manage the making of the mattresses
without assistance, any local maker will undertake
the work for a small additional cost per mattress.

Iron Bedsteads.—These should be 6 feet 4 inches
by 3 feet. They should be without curtains or val-
ances, and should have a bed-pull, by which the
patient can raise or move himself in bed.

Linen and Blankets.—It is essentially necessary
that these should be of good quality and finish.
Unbleached sheeting and pillow linen are decidedly
the most durable and economical, besides being
better adapted for hospital purposes than cotton.
Twilled cotton may be used with advantage for

children's sheets, but for no other purpose is it recommended.

An abundant supply of linen is very essential, and separate towels should be provided for each case. It may be of service to mention that the following is the average supply of linen allowed to each bed at the chief general hospitals :—

Allowance of Linen per Bed.

PATIENTS.

MEDICAL.			SURGICAL.		
Sheets	... ...	4	Sheets	...	5
Blankets	... ...	3	Blankets	...	3
Counterpane	...	1	Counterpane	...	1—2 over in each Ward.
Pillowcases	...	3	Pillowcases	...	3
Draw Sheet	...	1	Draw Sheets ...	...	4 Children's Ward, 6
Doctor's Towels (per			Doctor's Towels	...	12 per Ward.
Ward)	...	12			
Round ,, (per Ward)		6	Round ,,	...	6 ,,
Table Cloths ,,		4	Table Cloths ...	...	4 ,,
Tea ,, ,,		6	Tea ,, ...	...	2
Dusters ,,		6	Dusters	...	2
Shirts ,, ,,		6	Shirts (per Ward)	...	6
Finger Napkins (per			Finger Napkins (per		
Ward)	...	12	Ward)	...	12
Nightingale Cloak	...	1	Nightingale Cloak	...	1
Mattress or Bed (Flock			Mattress or Bed (Flock		
or Horsehair)	...	1	or Horsehair)	...	1—1 extra in each Ward.
Bolster	...	1	Bolster	...	1—1 extra for every 3 beds.
Feather Pillow ...	...	1	Feather Pillow	...	1—1 ,, ,, 2 ,,
Straw Palliasse...	...	1	Straw Palliasse	...	1

NURSES.			SERVANTS.		
Quilt	...	1 each and 2 over.	Quilt	...	1 each and 2 over.
Blankets	...	3 ,,	Blankets	...	3 ,,
Sheets	...	3 ,,	Sheets	...	3 ,,
Pillowcases ...	...	2 ,,	Pillowcases ...	...	2 ,,
Towels	...	4 ,,	Towels	...	4 ,,
Table Cloths	...	6 ,,	Rollers	...	3 ,,
Toilets	...	6 Oil Baize.	Tea Cloths ...	...	12
Mattress (Horsehair)		1 each.	Dusters	...	12
Bolster (Hair)	...	1 ,,	Table Cloths	...	6
Pillow (Feather)	...	1 ,,	Toilets ...	...	4 Oil Baize.
			Mattress (Horsehair)		1 each.
			Bolster (Hair)	...	1 ,,
			Pillow (Feather)	...	1 ,,

NOTE.—Size of Sheets, 3 yards × 2 yards 8 inches. Size of Pillowcases, 30 inches × 1 yard 4 inches. Size of Children's Sheets, 2 yards × 1 yard.

EXCEPTIONS.—1. Two Feather Pillows are allowed to each bed in the Male Accident Ward.
2. Extra Flock Beds are allowed in the Male Surgical Wards.
3. In the Children's Wards no extra Pillows or Bolsters are allowed.

21

Counterpanes.—It is a difficult matter to select a counterpane in all respects suitable for the purposes of a hospital. Durable, easily washable, of a bright and cheerful colour which will not fade in cleaning, —these are characteristics so difficult to meet in combination, that for many years, after extensive experience, it seemed to us impossible to obtain them anywhere. However, some advance has been made in this respect ; and Messrs. Boyd & Sons, of Jedburgh, N.B., have for years manufactured a counterpane specially for the use of hospitals. This counterpane is made of a beautiful material, soft as wool, yet easily washed and most durable. It is made in various shades, the favourite being blue grey with stripes at each side and end, about six inches wide. When placed upon the bed, it presents a very pleasing appearance, and some twenty-three years' experience has led us to the conclusion that this is the best kind of quilt extant. It may be obtained from the manufacturers in any size, at the cost of from 12s. to 15s.

Pictures in the Wards.—Much has been said for and against the propriety of ornamenting the walls with pictures, illuminated texts, and so forth. Our own view is, that in this, as in other things, the middle course is the best. We do not believe that, with proper care and management, pictures do, or can, in any way harbour infection. The chief thing to remember is, that the patients do not care for elaborate water-colours or magnificent landscapes. They like what they have been accustomed to at home—simple, homely sketches, of the cheapest character, with bright colouring and plenty of figures. The chromo-lithographs published by the *Illustrated London News* and *Graphic* are the most suitable. The present writer once introduced a number of these pictures into the wards of a hospital under his management, as they were the best he could then obtain for the purpose. Some fifteen months afterwards, Messrs. Graves, the well-known engravers, and a firm of chromo-lithographers, sent him some sixty of their most beautiful pictures. These were hung in the wards in place of the commoner productions. The patients were in a sad way at once, and one old man begged him to replace the original prints, for he declared " he could not sleep at nights until Dick "— alluding to a drawing of Dick Whittington and his cat—" had come back again." This story shows two things clearly. First, that these pictures do real good, by giving the patients something to look at and to think about when confined to their beds ; and, secondly, that the simplest and best known prints are much to be preferred to the most expensive pictures possible. If cleanliness is next to godliness, we are fully persuaded, from long observation, that it is equally true that cheerfulness in the sick is the highway to recovery.

The Walls of the Wards.—Care must be taken, as we have insisted upon elsewhere, that every piece of paper is taken off the walls, and that the whole cottage, if about to be adapted to the purpose of a village hospital, is well cleaned from top to bottom. It will be found an excellent plan, when the walls are old, to have them repapered, and then to varnish the surface of the paper with two coats of the best copal varnish. This will give a very pleasant appearance to the room, and it will enable the nurse to thoroughly cleanse the walls, as often as may be thought necessary, since she will be able to sponge, scour, or mop the whole of the surface without injury to the paper, or damage to the wall, and thus the utmost cleanliness will be ensured. A far more important end will also be gained. The paper, thus varnished, will present a hard, dry, non-absorbent, and practically impervious surface, thus ensuring the utmost security against an outbreak of pyæmia or erysipelas. This plan has been tried in several hospitals with excellent results ; and in proof of the durability of such a proceeding, it may be mentioned, that one coat of varnish has been found to be sufficient to preserve the wall and paper intact for nearly ten years. There is no doubt, as the present writer can prove from actual experience, that where varnish has been used in hospital wards which have previously been painted with four coats of good oil paint, a surface is presented as smooth as the finest Parian, as impervious as adamant, and as durable as it is possible to make it. Where this plan has been tried, the general sanitary and hygienic conditions of a hospital ward are rendered almost perfectly secure.

Lint and Medical Sundries.—These can all be obtained, of excellent quality, and at a greatly reduced cost, from a large number of firms, the names of whom will be found in *The Hospital* newspaper, which should be kept filed in the committee room, as it will be found the most up-to-date source of information upon such matters.

Patients' Necessaries.—Below we give a list of articles of which the patient stands in daily need, and for the supply of which some arrangement must be made. As some of these are liable to breakage, and all of them to be mislaid or lost, an arrangement must be devised by which the cost of any such loss shall fall on the patient. Every properly administered hospital should provide these articles for the patient, and make him or his friends pay a certain deposit on them, to be returned when he leaves the hospital, after a deduction for any losses that may have occurred, according to the prices on the list ; for example :—

	s.	d.
Small Earthenware Teapot,	0	6
Cup and Saucer,	0	3
Two or three Plates,	0	4
Basin,	0	2
Knife and Fork,	1	0
Tablespoon,	0	3
Teaspoon,	0	2
Total,	2	8

A cruet-stand and salt-cellar should be provided for the use of each ward by the hospital. Very pretty cruet-stands can be made in fretwork. Should there be a lady or gentleman in the parish who does such work, two stands might be given to the hospital at almost a nominal cost. They are easily fitted up with bottles at any glass shop.

Each patient will, of course, be expected to bring with him such things as he may require for his personal use and comfort, as razor, brush and comb, etc.

Lockers.—Lockers to stand by each bedside, and to stow away the above-named articles, together with the allowance of bread, butter, cheese, etc., for the day, will also be requisite. Should there be no such provision, these things will probably be found heaped together in an untidy mess beneath the pillows. The different kinds in use are so numerous, that it is impossible to specify any one as particularly good. In fact, a particularly good locker is much needed, and would be a very useful invention. As a rule, they are made to stand on the floor, and are placed too much out of the reach of the patient, who has to lean over considerably to get things out of them,—not always an easy matter in cases of fractured limbs. As lockers may be made by the village carpenter, we give the chief points in their construction, so that it will be easy for each hospital to select its own plan. They should be raised nearly to the level of the bed, that they may be of easy access to the patient. For the same reason, they should open neither at the top, nor in front, but on the side nearest to the patient, by a sliding door, provided with a simple catch. The interior must be of sufficient size to hold the things

comfortably, and be fitted up with one or more
shelves. It ought also, if thus made, to be more
deep than wide. If thought preferable, it would be
easy to devise double lockers after this pattern with
a partition down the centre, one thus serving for two
beds. This locker has the advantage of a larger
top, which might be used as a table, on which the
patient could place anything, play games, etc. If
made to open at the top, as is frequently the case,
it is necessary to continually disarrange the things
placed there, and one of the most important uses of
the locker is thus lost. A roller should also be
provided behind, for the patient to hang his towel
upon.

Another useful arrangement in vogue at some
hospitals is to have a small *bracket* fixed to the wall
above each patient's bed, and out of his reach, on
which the medicine bottles may be placed. It is
thus more easy for the nurse to go round and administer the medicine, and there is less risk of a mistake
being made. In the case of a delirious or troublesome patient, the bottles must, of course, be removed.

Headings for Beds, etc.—We give over leaf an example
on a small scale of a useful prescription paper to be used
for each case, together with a second sheet, which is
useful in long-standing cases, when the first has been
filled. The use of these in every instance, and the desirability of their being accurately filled in, must be
particularly urged. By this means, if they are afterwards arranged in order in a portfolio, a case-book of

Cottage Hospital.   No.

Name,	History,
Age,	
Occupation,	
Address,	
Disease,	
Date of Admission,	
„ Discharge,	
Result,	Medical Attendant,

Date.	Prescriptions.	Diet.	Remarks.

Cottage Hospital.

Name,   Case No.   *(continued)*.

Date.	Prescriptions.	Diet.	Remarks.

PRESCRIPTION PAPER.

every patient admitted is easily kept, and reference can readily be made to the cases for future guidance. These papers should be placed by the bed to which they refer. One plan is to hang them in a frame, on a nail fixed in the wall. The frames are sometimes made of tin, with a narrow bar fixed to the top through which the paper slips, the bottom being turned up and soldered down at the sides, to form a ledge for it to rest upon. Others are made of pasteboard, with a piece of leather fixed to each corner, through which the tips of the paper may be slipped. Another pattern, and a more simple one, is in use at the Homerton Fever Hospital. It consists simply of a board about $\frac{1}{4}$ inch in thickness, and rather larger than the paper, which is fixed to it by means of two drawing pins at the upper corners. The lockers in this hospital are surrounded, except in front, by a ledge (also a useful improvement), and the top of the locker projects behind about half an inch. On this projection the board rests, whilst on a level with the top of the ledge runs a bar or strip of wood, fixed about $\frac{1}{4}$ inch from the ledge by means of two projections at the sides. The board is easily slipped through this, on to the projecting top, where it is kept firmly in its place.

Ward Furniture.—Drawers or cupboards to contain the patients' clothes, when not in use, must be provided. It will be best to place these outside the ward, and in charge of the nurse. By this means, a patient will not be able to get up and dress without the nurse's knowledge. There should be a rule, too, that at night every patient, after undressing, must deliver his clothes to the nurse, who will place them each in their own separate drawer. The filthy practice of placing the clothes under the mattress for the night cannot be sufficiently condemned.

A good sized table in the middle of the ward looks homely, and will prove useful, even if such patients as are able dine together in the kitchen or convalescent room, so that it is not required for purpose of meals.

Movable Closet and Bath.—A movable earth-closet in each ward will be the most efficacious means for the use of such patients as are not able to walk to the closet. A movable bath, which can be wheeled to each bedside, will, if funds are ample, prove a great boon in many cases.

Hot-water Plates. — Hot-water plates, or dishes with covers, to contain the diets on their transit from the kitchen to such patients as are confined to bed, will be of service. The use of these can be limited to such cases, but they will prove generally serviceable, as patients do not always feel inclined to take their diet at the moment it is brought to them.

Screens.—Two folding screens would conduce to decency in many cases, when, for instance, it is desirable to hide the bed whilst a patient is being examined by the surgeon, or to spare the feelings of the others when one is *in extremis*. These screens can easily be made an ornament to the ward, and a source of interest to the patients in the following manner.

Get a carpenter to put together some light frames of a suitable size, nail canvas round them, and afterwards hinge three or four frames together. On these frames may then be pasted a number of pictures, cut out of the *Illustrated London News*, *Graphic*, or other papers, so as to completely cover them. One of these screens, made with coloured scraps, is really a very great ornament to any house, but the scraps need not of necessity be coloured. It will in most cases be easy to get one of the ladies of the parish, who takes an interest in the hospital, to undertake this task. If approved, the screen may afterwards receive two coats of a solution of isinglass, and be then varnished. The isinglass is to prevent the action of the varnish on the colours.

Bed Rests.—The bed rests of the present day are often of a most complicated description. We suggest a simple one that can be made by any carpenter. The hinder part is fastened to the front by two hinges at the top. At the bottom are two pieces of window-blind cord, passing from the large transverse piece in front to the piece behind, to prevent  them from slipping too far asunder. A useful and inexpensive bed rest modelled in accordance with our suggestion, is made of galvanised iron or cane, and is strong, light, and portable. The sketch on p. 315 will give an idea of it. To use this rest the bolster and pillows must be removed, and the rest firmly placed on the mattress. A pillow is then placed over the framework, against which the patient leans. This rest will be found to answer its purpose quite as well as the more complicated contrivances.

A chair makes a very efficient bed rest in an emergency. The legs of the chair are to be turned upwards and the seat backwards, so that the upper part of the back of the chair slopes gradually under the back of the patient, who rests against it on pillows.

Arm Slings.—In all fractures of the upper limb, with a single exception of that of the olecranon (the point of the elbow), great assistance is derived from supporting the limb in a sling, with the elbow bent at a right angle. The common practice of merely passing the sling under the hand or forearm, or both, is very inefficient. It is the elbow which above all requires supporting to carry the weight of the arm ; and this will be found to be quite as true if the injury be to one of the soft parts or of the bones of the thumb or of the humerus.

To make such a sling for the right arm of an adult, take a piece of black alpaca or other suitable material 29½ inches wide, and 35 inches long ; fold it corner-wise with the smallest triangle inwards ; the outer one will thus have about five inches projecting beyond the half square. The hand rests in the base of the triangle, the opposite angle is at the elbow, which is efficiently supported by the projecting portion of the outer layer being carried round the lower end of the arm, where it must be pinned or stitched to the inner part of the sling. The two other extremities are carried upwards, on the outer and inner aspect of the forearm respectively, and are secured by a strap of the same material at the back of the neck.

Surgical Hammock.—To lift a patient is often a matter of difficulty, and a cause of bodily injury to the nurse unless efficiently assisted. Many patients may help themselves considerably by the old plan of a rope or long towel fastened to the post or cross-bar of the bed. When a patient is heavy and very helpless, he will be found much more manageable on a single than on a double bed, so that attendants on either side can assist each other. They can do so efficiently in lifting the patient by firmly grasping each other's hands under the thighs and in the middle of the back. A roller towel passed under a patient's buttocks and grasped by an assistant on each side, is useful in lifting. Much greater power may be exerted by slinging each end of the roller towel over the attendant's neck.

The woodcut on next page illustrates a surgical hammock or lift which Messrs. Salt & Son of Birmingham have constructed at the late Mr. Sampson Gamgee's suggestion. This apparatus can be put together over the bed, so that one person can raise the heaviest patient, retain him in an easy swing for any length of time, and lower him on to the bed whenever required without jerk or effort.

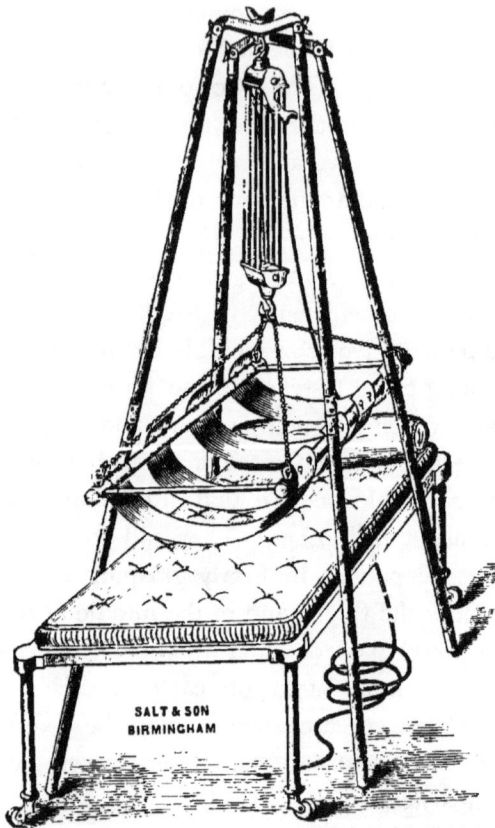

Under the patient, and next to the night-dress, are to be placed some soft wide pieces of webbing, of sufficient length to allow scope for the movements of the body within the rods lying on each side of it. These rods are connected with the ends of the pieces of webbing so as to secure strength and to be easily removable. When it is necessary to lift the patient, a stout four-legged derrick is adjusted with two of its legs on each side of the bed, having a powerful self-stopping pulley suspended by a hook to the point where the four legs converge and unite. The inferior block of the pulley is armed with a hook, which grasps the chains connected with the rods lying on each side of the patient. To prevent these rods approximating too closely and from squeezing the body uncomfortably, transverse bars are fitted into sockets at their extremities.

The body is thus hung in a hammock, which can be raised or lowered by one attendant very gradually, and with the certainty that when the cord is drawn to one side the pulley will catch instantly and immovably, and run smoothly again, as soon as the cord is drawn straight in the middle line. Thus suspended, a patient can rest comfortably for any length of time, the bed can be made, or any surgical dressing can be applied with ease and comfort.

When the patient is lowered, the derrick, frame, pulleys, and subsidiary adjuncts are removed, while the webbing, straps, and longitudinal rods may be left in readiness for future use. Thus, in hospitals, the elevating apparatus may be made available for many patients, the separate webbing straps alone being required for each bed. For greater convenience the poles unscrew in their centres, and can be separated by means of thumb screws from the cruciform connection which sustains the pulleys.[1]

Easy Chairs.—The most comfortable kind of easy chair, especially for cases of fractured legs, is a low one, made long in the seat and sloping somewhat at the back. Foot rests of about the same height as the seat of the chair should also be provided for these cases, so that the leg may not hang down, but be supported at the same level as the hips. This is a great point, as it prevents the leg from swelling, but it is too often overlooked. These chairs may be purchased for about £4 each.

Filter.—A filter for water will in almost every case be an absolute necessity. There are so many good patterns that it is not easy to name the best; but Crease's carbolite filters, as used in the Navy, are good, and perhaps the safest. They may be procured from F. Braby & Co., Limited, 362 to 364 Euston Road, London, N.W.

A *wash-hand basin and jug* with a towel should be placed in each ward for the use of the surgeon in his rounds.

Book Shelves.—The homely and cheap book shelves, made of three pieces of wood of differing sizes, the largest piece being placed at the bottom, and all held together by pieces of window-blind cord, passing

---

[1] *On Surgical Swings and Pulleys as Aids to Rest and Motion*, by Sampson Gamgee. F.R.S.E., surgeon to the Queen's Hospital, Birmingham.

through the four corners of each, and holding them
there by means of a knot, the ends of the cord being
tied together at the top, will prove as serviceable as
any. If preferred, pieces of red cloth, cut out in any
pattern, may afterwards be nailed round the edges of
the shelves, and will much improve their appearance.

Ambulances.—Many cottage hospitals are now supplied with an ambulance for the removal of patients
to the hospital. *The Hospitals Association* litter, used
for the street ambulance service in London, and
made by Carter, 47 Holborn Viaduct, is excellent.
Dr. Swete has invented a very useful one, and it has
been variously modified for different cases by
others. It is best made to order from the local
coachbuilder.

Dr. Swete gives the following account of his ambulance, which he states can be made for twenty guineas,
and leave a fair profit at the same time for the local
coachbuilder :—

" The body is like a skeleton hearse, without bottom
or sides, the panels at the side being filled with glass
shutters or curtains of vulcanised india-rubber cloth,
capable of being drawn back opposite a wound, etc.
The back of the carriage opens as a door. In the
bottom are three rollers, on which runs a wooden
tray, having placed in it a mattress covered with vulcanised cloth (or in infectious cases, straw which may
at once be burnt). This tray has two strong iron
handles at each end, and one at each side, so that it
can be conveniently carried up a narrow staircase.

22

The vehicle is hung by four elliptic springs on four wheels, and is capable of being turned in its own length. It has a driving box, and shafts for a pony or donkey, which can be obtained in nearly every village, are strapped on the roof, with a handle to draw it by hand. On each side is a rail, so that any friend or attendant, whilst watching the patient, and walking with the hand resting on the rail, will give considerable assistance in propelling the carriage. A lamp is fitted outside and inside the carriage. No ornamental work exists. The wood is ash, with deal panels ; the wheels are ash, with turned spokes. There is neither paint nor putty used, the whole being varnished inside and out, and the iron-work bronzed. A spanner is attached to the driving box, ready for use to change the handle for the shafts. The name of the hospital or union to which the ambulance is attached, is painted on an ornamental ribbon or scroll, on the door. Its dimensions are 7 feet 6 inches, by 3 feet 9 inches, the height 4 feet 9 inches. The tray in which the mattress is placed is 6 feet by 2 feet. The total weight is under three cwt. The appearance is not inelegant, and the carriage would not look *outré* in the entrance hall of a hospital or union-house."

The mode of use is as follows :—"The carriage being brought near the place of accident, or where the sufferer lies, the tray is taken out and carried to him. This may be taken down a railway embankment, over a wall or locked gate, or up a narrow staircase. The patient, being placed on the bed, is carried to the ambulance, and in the same way from the ambulance to the hospital bed."

With reference to this ambulance, Dr. H. Franklin Parsons of Goole wrote to the *Sanitary Record* of June 17, 1876, as follows :—"The Goole Local Board have recently had an ambulance built, which answers well. The idea of the vehicle was taken from that figured in Dr. Swete's *Handbook of Cottage Hospitals*, but closed sides with windows were, for infectious cases, considered preferable to the open body with curtains recommended by Dr. Swete. The body is something like that of an omnibus, 7 feet long, 3 feet 9 inches broad, and 4 feet high, opening with a door at the back. The sides are boarded for the lower two-thirds, with sash windows above, two of which will let down. In the bottom is a wooden tray for the patient to lie upon, 6 feet by 2 feet, sliding on rollers, and with handles at the ends and sides, so that it can be lifted out and carried upstairs. The tray lies along one side of the van ; and being narrower than the body, the remainder of the breadth forms a gangway for the person who accompanies the patient, and who sits on a stool. The under-carriage was that of a second-hand phaeton ; the wheels are respectively 2 feet 6 inches and 3 feet 6 inches in diameter, with patent axles. There is a driving box in front on the top of the van. The whole machine is very light, so that one horse can easily draw it. It was made by a local coachbuilder, and cost £22."

Miscellaneous Articles.—Below will be found a list of

articles required for use in the hospital.   They admit
of no particular description.   In looking through the
report of the different cottage hospitals we find that
in most cases these or many of them have been given
as presents to the institution, and probably this will be
the rule.   Many of these things are obtained by persons for cases of illness occurring in their families, and,
when they have no further use for them, they are very
pleased to get rid of them for such a useful purpose.

Foot Warmers.	Nightingales or Garibaldis.
Water Bed.	Commodes and Bed Pans.
Air Cushions.	Dressing Gowns.
Cradles.	Slippers.
Crutches.	Foot Stools.
Slings.	Books.
Cushions.	Scrap Books.
Sand Bags.	Old Linen Rag.
Feeders.	Inhalers.

Bags and pads of great variety may be made by the
nurse as means of applying pressure, steadying limbs,
or absorbing discharges.

Sand bags are best made with bed-ticking and fine
sand or earth well dried by putting in the oven or by
exposure to the fire.   These bags are usually round
and long, but their shape and weight must vary according to the requirements of particular cases.

Shot bags are made with bed-ticking or wash-leather
and small shot, which may be prevented from rolling
about by rough quilting.   The square is a most useful shape for shot bags to allow of their lying comfortably on parts requiring gentle steady pressure.

Oakum and tenax (Southall's) *pads* should be made by well teasing out the material, and enclosing it in tarlatan or thin book muslin bags, roughly stitched.

Sawdust pads may be made with the same muslin and roughly quilted. All sawdust is absorbent, and very useful for surgical purposes. As Surgeon-Major Porter first suggested, pitch pine sawdust is preferable. It possesses undoubted antiseptic properties, and its odour is very agreeable. Indeed, terebene is now proved to be a useful and pleasant disinfectant.

Tow pads, made like the preceding, are also very convenient and comfortable, as absorbing and smoothly compressing agents.

The patent safety or nursery pins should always be preferred to common pins; they hold faster, and cannot wound.

Hot-water tins can be purchased from most instrument makers and furnishing ironmongers. When the special article is not at hand, ordinary bottles may be used; stone ones are preferable to glass. They should be wrapped in flannel, to regulate their temperature and to prevent the scalding of the patients.

A feeding cup with a nicely curved spout, all the better if at right angles with the handle, is an article that no nurse should be without.

Bed pans should be of the slipper shape; a loose flannel covering for the pan gives great comfort. In winter the nurse should always keep the bed pan near the fire. Before giving it to the patient, and immediately after use, some of the disinfecting powders sold in dredging boxes should be sprinkled in the pan.

In closing this chapter, it is necessary to remark that these things are not all absolutely necessary, at any rate on first starting. It is, however, urged, that it will be wiser not to put by surplus money as an endowment, till most of them have been provided, but to expend any surplus in making the hospital more complete as regards its fittings. The list may also prove serviceable as a reference for such persons as are minded to make a present to a hospital, since they can thus easily note in what articles the institution may be deficient.

CHAPTER XII.

COTTAGE HOSPITALS IN THE UNITED
STATES OF AMERICA.

Commencement and growth of the movement in the New England States—Lack of accommodation elsewhere in the United States—Baneful influence of politics on management of municipal hospitals—Superiority of those administered on the voluntary principle—Evils of indiscriminate medical relief—The constitution of cottage hospitals in the United States—Their financial system—Patients' payments—Municipal grants—Endowment of beds—Medical and nursing arrangements—Training of nurses in cottage hospitals—The matron's position—Register for nurses at cottage hospitals—Isolation of contagious diseases—Maternity cottages—Convalescent cottages—Hospital cottages for children—Cottage hospital construction—Pay wards and their advantage in the training of nurses.

IN the second edition of this book, which was published in 1880, we urged that so useful and successful an institution as the cottage hospital could not long exist in England without being adopted on the other side of the Atlantic. At that date the movement was just obtaining a foothold in America. Circulars of inquiry were about that time sent by the State Board of Health of Massachusetts to their correspondents in every town of 10,000 inhabitants and upwards ; and in every case where no hospital existed—and this was the rule—the correspondent

(327)

stated that in his opinion such an institution was needed. New England has always taken the lead when a display of unusual intelligence has been called for on the part of the citizens of the United States, and we are not surprised therefore to find that the action of the Massachusetts State Board of Health made the establishment of cottage hospitals a public question, which was enthusiastically taken up throughout the State. In 1880, whereas Boston possessed fifteen hospitals, with accommodation for over 1000 patients, there were in the rest of the State (exclusive of institutions for the insane and those belonging to the naval and marine services) only ten hospitals, with an aggregate of 240 beds. At the present time it will be seen from the list of cottage hospitals given on p. 363, from which we have been enabled to obtain reports through the courtesy and kindness of Dr. Rowe, of the Boston City Hospital, that the movement has prospered greatly. Indeed, there is reason to believe that something like a majority of the towns with 10,000 inhabitants or upwards have now at any rate one small hospital for the sick.

Before the movement arose in New England the only available places for emergency cases and accidents were the police station and the almshouse, that is, an inferior sort of workhouse, the very name of which was repugnant to the finer feelings of the majority of people, irrespective of their station in life. The almshouse possessed no hospital accommodation in any proper sense of the term ; it was often already over-crowded, and the material for intelligent nurses could not be found among the inmates, who were either old or enfeebled, or who represented the waste products of the population. Writing of the condition of affairs at Waltham, Mass., Dr. Worcester, so recently as March, 1894, states : " For the stranger within our gates, for the traveller fallen by the wayside, for the farm-hand or for the servant girl when stricken by disease or maimed by accident, there were the alternatives of the Massachusetts General Hospital and the Town Almshouse. Often there was room only in the latter. It seems incredible that such was the condition until so recently. When I recall the terrible state of things in the old almshouses, where the poor fever patients were carried to be nursed by the paupers in their dotage, or when I remember how others sick or wounded were carried in common railway hacks ten long miles over the roads to Boston, I feel again the burning sense of shame. Yet Christians were not more heartless then than now. They were waiting for their pathway of duty to open."

This description by Dr. Worcester of the state of affairs at Waltham up to a few years ago may, we understand, be taken as a fairly accurate description of the state of affairs which prevails in very many portions of the United States at the present time. The cause seems to be that the population as a whole have a dislike to enter a public hospital, most of these institutions being subsidised, if they are not entirely supported, by the municipal authorities out of the rates. Politics enter largely into municipal affairs, and as a consequence the management of many municipal hospitals is discreditable, and well calculated to inspire the population with fear and dread for the consequences which may follow the admission of a patient to one of these institutions. Such being the system, however, whenever the need for hospital accommodation has arisen in a community in the United States, people have looked to the municipality to provide the money, or a considerable portion of it, to pay the cost of site, buildings, and maintenance.

Where those responsible for the rates have felt reluctant to make grants for hospital purposes, the people after due hesitation have very frequently taken the matter into their own hands, and subscribed the funds for a hospital established upon the voluntary principle. Wherever this has taken place the results have been satisfactory; and we are disposed to think that as the blood of the martyrs proved to be the seed of the Church, so the sufferings of the few for the want of adequate hospital accommodation in the past have saved the population from the evils of municipal hospitals, and secured for them institutions founded by loving hearts, maintained by the free-will offerings of the people, and administered with remarkable efficiency and success. No doubt the development of cottage hospitals, which is mainly confined at the present time to New England, will gradually spread to other States until the whole country will be provided with adequate accommodation for the sick of all classes.

In 1891, and again in 1893, we made a careful inspection of a great number of the existing hospitals of the United States. We found, speaking generally, that the hospitals administered upon the voluntary principle were efficiently conducted, whereas those in the hands of politicians were unsatisfactory, and in no sense a credit to the country. We further noticed with regret that everywhere there seemed to be a growing tendency to increase the number of free beds, and thereby to imperil the independence of the people who use hospitals. We hope that all who are interested in the efficiency of hospitals in the United States will set their faces like a rock against any large increase in the number of free beds, and that they will determine to maintain the principle of paying wards and pay beds; because this system, which had its origin in America, is the true principle upon which medical relief can be given to the whole population to an adequate extent, without any danger of serious abuse of any kind.

So far we have spoken mainly of the condition of affairs in urban districts. Here, as Dr. Presbury has pointed out, " the scanty means of many persons force them to live in crowded and unwholesome tenements, under conditions favourable to the development and spread of disease, and most unfavourable to its treatment. Boarding houses, which are suitable to those who are well, are not adapted to or equipped for

the care of the sick ; and the boarder who finds himself confined to his bed by sickness is likely to feel the
need of home comforts and attention, and his peace
of mind is not increased when he finds he may be a
source of inconvenience and possibly of danger to his
fellow-boarders. This distress is all the greater to one
who has suffered severe injury, and is in need of
surgical attention and constant care. Hence prudence
dictates that a State increasing in population, and connected so closely with several manufacturing places,
should be provided with convenient and properly-
constructed hospitals." These views have commended
themselves to the majority of town residents in New
England ; and that, no doubt, accounts for the large
and increasing number of hospitals which have been
established during the last fifteen years.

In the earlier days in towns, and at the present time
in many rural districts, the conditions concerning the
sick and injured are much as follows :—In such districts everybody knows everybody else, and the sick
have been cared for by kind neighbours. Dr.
Worcester states that even in the families of the most
well-to-do it was customary to accept the kindly
services of night watchers, who came by turns to the
relief of the stricken family. In this case some one
neighbour is especially depended upon, and there
exists a genuine neighbourly helpfulness amongst the
population. Neighbourliness, including the good old
custom of watcher for the sick, is, however, fast going
out of fashion ; and as the need for cottage hospitals

in the rural districts increases, no doubt the example
of the towns will be followed, and these institutions
will gradually be established, first in New England
and ultimately throughout the United States.

As the cottage hospital movement is now making
such considerable progress in the United States, it
may be useful to give a brief account of the steps
taken to provide against the evils of indiscriminate
medical relief, to secure justice to the medical profession, hygienic conditions, economy, skilled nursing,
and the introduction of certain special but necessary
features, including an adequate provision for fever
patients, cases of midwifery, and convalescents.

The Constitution.—It may be useful at the outset
to give a brief outline of the constitution of cottage
hospitals in the United States. The cottage hospital
is generally incorporated under the laws of the State
in which it is situated. The constitution or charter
provides that certain persons whose names are
given, and others who may hereafter contribute sums
varying from $1 to $10 a year to the funds, shall
be members of the corporation of the hospital. The
election of life members from donors of larger sums
is also provided for in some cases. The affairs of the
hospital are placed in the hands of from 25 to 30
Trustees elected from the members. The Trustees have
control over the whole management of the hospital,
but they delegate some of their powers to committees.
Thus, there is usually an Audit Committee of two or
three, charged with the duty of managing and audit-

ing the accounts ; a Finance Committee, of from three
to five members, who collect funds for the hospital ;
and an Executive Committee, to whom is entrusted
the general management of the hospital, including the
appointment of the medical and other members of the
staff, and the admission of patients and other necessary details relating to the general conduct of its
affairs. The Trustees generally meet once a quarter
to receive the reports of the various committees, the
Executive Committee holding monthly or bi-monthly
meetings according to circumstances. Those Trustees
who are on the medical staff are expected to visit the
hospital at least once a week, according to a rota
settled by themselves. The members meet annually
on seven days' previous notice, and special meetings
may be called by the President or by a requisition
signed by three or more members. The principal
officers and the Trustees are elected by the members
in annual meeting.

The Trustees, who usually consist of ladies and
gentlemen, of whom one-third retire every year,
but are eligible for re-election, are sometimes
called the Board of Directors, or Board of Managers.
They at their first meeting elect a President,
Vice-President, and the various committees, the most
important of which is the Executive Committee,
on which the medical staff is usually represented.
The religious ministrations to the patients are provided for by the admission of any minister whom a
particular patient desires to see. The rules for the

government of the institution, when approved by the
members, may be amended by vote of two-thirds of
the members present, provided due notice has been
given in advance. There is a rule with regard to
operations which we believe is confined to hospitals
in the United States, which permits a private patient
being brought into the hospital for an operation on
payment of not less than $5, provided he is removed
immediately after recovery from the anæsthetic. It
is usual for patients applying for admission to present
to the matron a card signed by a physician practising in
the town or district from which he comes. The patients'
friends are admitted to see them on two or three days
a week, generally from two to four in the afternoon.

Finance.—Finance is necessarily the keystone of
the whole system of cottage hospitals. The difficulties
of providing the necessary money to purchase sites,
erect buildings, and maintain the hospitals, have been
in the United States as great as, perhaps greater than,
in Great Britain. So far as we have been able to discover, the various religious bodies, which often provide
large general hospitals in the great towns of America,
have so far held aloof from the cottage hospital movement. When, as in the case of the Morton Hospital
at Taunton, Mass., the strong and urgent recommendations of the City Physician that the municipality should establish a hospital, failed to convince
the City Government of the need of action to the
extent of appropriating the necessary money, the
people took the matter into their own hands; the

Bristol North District of the Massachusetts Medical Society took up the matter, and Dr. Hewens suggested that a company should be started by voluntary subscription, to be ultimately incorporated. No practical result immediately followed. Later on, some of the citizens started a movement to secure the incorporation of a certain number of subscribers, each of whom pledged himself to give at least $10 a year for the next five successive years, for the purpose of establishing a hospital at Taunton. In this way 125 subscribers, in addition to the seventeen societies, were induced to pledge themselves to this extent, a meeting was held, officers were appointed, and bye-laws approved. So much public interest was aroused that Miss Susan J. Kimball of Boston presented the committee with an estate, and the hospital was successfully incorporated on the 20th June, 1888. Public interest in the movement grew, gifts of materials and money flowed in, and the societies connected with Protestant churches supplied the bed linen and other useful materials.

The history of other hospitals is pretty much the same. Each one has had to pass through the experimental stage, during which the community had to be convinced that a hospital was needed, and then their support and sympathy have been secured. The most popular method seems to be the establishment of a hospital and an association of ladies, the object of such an association being to furnish the hospital and assist in its maintenance in such ways as

shall be deemed most suitable. It is controlled by a
board of directors elected in annual meeting, which
selects suitable persons in the various communities to
serve as soliciting committees. Each community from
which the patients are received is canvassed, and each
sends subscriptions to the fund for the maintenance of
the hospital. The association labour to secure a permanent endowment for their hospital and to induce
people to remember the institution in their wills, and
to give considerable sums by special gifts or bequests
in order to provide for new features.

One fruitful source of income is patients' payments,
which usually are equal to at least 25 per cent. of the
total expenditure. The municipality frequently makes
a grant of from one-eighth to one-quarter of the total
expenditure; another quarter is derived from donations, annual subscriptions, and Hospital Sunday
grants; while the remainder is provided by the income
of special funds, and the profit from supplying nurses
to private houses. The movement is too young to
have produced many legacies, but this item is growing
in importance and amount. The average weekly cost
per patient shows enormous differences. Thus, at the
Lynn Hospital with about 40 beds, of which 28 are
constantly occupied, in 1894 the average weekly cost
per patient was $8·50; whilst at the William W. Backus
Hospital at Norwich, Connecticut, containing 63 beds,
15 of which were constantly occupied in 1894, the
average weekly cost per patient amounted to $28·76.
On the other hand, it is stated in the report of the

23

Elliott Hospital managers that the cost per week per patient in nineteen New England hospitals averages $12·61. Roughly speaking, we find that the Waltham Hospital (20 beds, of which 17 were constantly occupied) expended in 1894 $7237 on maintenance, $1711 on nursing services, $2843 on salaries and wages; about $12,200 altogether, or $2 per day per bed. It may be interesting to add that many people seem to welcome the opportunity of endowing beds. $300 per annum seems to be the cost of furnishing one bed, and various Sunday schools combine to establish a children's free bed fund in many of the hospitals. Most of the rules provide for the endowment of beds for one gift of $5000, or by an annual contribution of $300, the estimated cost. The aim seems to be to so organise each hospital that one-half of the beds shall be free by endowment, and the remaining half available for paying patients.

The payments made by the in-patients vary from $2·75 for men and $1·75 for women per week at one hospital, and from $5 to $7 per week at others for patients in the ordinary wards, to $30 per week in a private ward. The charges for out-patients, exclusive of fees for professional attendance, seem to be about $2 for treatment with ether, $1 for treatment without ether, and $0·25 for each subsequent treatment.

The income derived from the nurses' services at the Newton Hospital is about $3000 per annum, and patients' payments amount to $6500, the total income being about $30,000.

Medical and Nursing Departments.—The rules of the American cottage hospitals usually provide for the appointment of a hospital staff and an auxiliary staff. The hospital staff consists of physicians enrolled to attend upon the service of the hospital, their number and the circumstances of their appointment being subject to local conditions. The auxiliary staff consists of physicians who are expected to render temporary services during the enforced absence of members of the hospital staff. The preference is given to members of the auxiliary staff in filling up vacancies on the hospital staff. The hospital staff treat charity patients, and patients not otherwise provided for, without compensation, except in cases of variola.

Physicians resident in the town where the hospital is situated are permitted, with the approval of two medical members of the Executive Committee, to attend inmates of the hospital who pay special rates for board, and who have been admitted there in the usual way. The attending physicians in severe cases are required to call in one or more of the hospital staff in consultation, but otherwise they may call in in consultation the physicians of their school, and with their consent any other physicians, provided it is done without expense to the hospital. It seems rather funny to our English ideas to find that patients may elect, upon their admission to a hospital, the school of medicine by which they shall be treated, and that when no preference is expressed the matron assigns patients in alternate order to the two schools—"the

medical and surgical cases being considered in
classes," whatever that may mean. When the two
schools receive and treat their patients in the same
room, "the respective patients shall be located upon
the opposite sides of the ward, if possible, and each
division so designated as to give information and
avoid confusion." Hence it is not uncommon to find
a notice suspended on one side of the fireplace which
states that all patients are treated on the homeopathic system "at this end of the ward," and on the
other side of the flue a notice to the effect that patients
" from this point " will be treated allopathically.

It seems to be the almost universal practice to have
a training school for nurses attached to each cottage
hospital. The justification for this method is said to
be that skilled nursing is growing to be an occupation
which offers to suitable women a means of livelihood,
that there is a steady demand for more nurses in the
hospitals and in each city generally, and that the
nurses, when trained, add to the income of the
hospitals by their services outside. No details are
forthcoming of the wages paid to the nurses, but the
profits to the hospitals seem to be considerable, seeing
that, as we have stated above, the income brought in
by nurses to the Newton Hospital during the year
amounts to about £600. Another reason which seems
to influence the managers in sanctioning the establishment of a school for nurses is that experience has shown
that it is cheaper to train their own nurses than to hire
them ready-made or from other institutions. It is

further contended that the smallness of the institution, and the consequent eagerness of all connected with it to promote efficiency, furnish unusual facilities for the instruction of nurses. The matrons and physicians are enthusiastic and competent, and the plan keeps up a succession of persons able to do the work, and ultimately sufficient in number to supply the demand outside the hospital. Our information leads us to conclude that this view is not entirely shared by the Association of Superintendents of Nurse Training Schools in the United States. Representative members of this body have more than once declared, that in organising an association of superintendents great care is necessary to avoid the admission of members attached to institutions where the training is inadequate or inefficient—that is, we presume, in its extent and completeness when compared with the larger facilities for instruction afforded by the best managed general hospitals throughout the country.

In cottage hospitals in the United States the matron is usually the most important resident officer. In some cases she has the assistance of an honorary medical superintendent ; but her responsibilities are clearly defined, as will be gathered from the following outline taken from the rules : —

The matron must be an educated and skilful nurse, capable of taking charge of the hospital under the direction of the Executive Committee, and of instructing pupils in nursing. She must have had a training in hospital work, at least equivalent to that furnished

by the hospital training schools. She shall have the
whole responsibility for the conduct of the hospital
under the direction of the Executive Committee,
according to such rules as she may find it necessary to
prescribe, and as may be approved by the Executive
Committee. She shall act as superintendent of the
training school and registrar of the directory for
nurses (*i.e.*, the nursing institution or home from which
nurses are supplied to private cases). She shall have
the appointment of the cook and other female
servants, the pupil nurses, and the assistant nurses,
subject to the approval of the Executive Committee,
and the power to discharge them. She shall not
expend any money or contract any debt without the
authority of the Executive Committee. She shall keep
a record in detail of all money received and disbursed
by her, which shall be open at all times to the inspection of any member of the Executive Committee, and
shall render quarterly to the said committee a full
account of all her receipts and expenditures. She
shall keep an accurate account of bedding, table and
other furniture, shall take charge of all money or other
property not in use, belonging to patients, and keep a
record of the same, and shall deliver it to the patients
upon their leaving the hospital, taking a receipt for it.
She shall, when a patient dies, furnish the Executive
Committee with an inventory of all the effects, and the
committee shall direct their disposal.

Many cottage hospitals have a register for nurses, so
that they can supply families and physicians desirous

of securing the services of nurses for patients in private houses. Each nurse connected with the registry pays a yearly fee of $5. A person obtaining a nurse from the registry pays a fee of $1 at the time when the engagement is made, which fee may be increased at the discretion of the management. The managers reserve the right to remove the name of any nurse from the register, at their discretion, for cause. A register of graduate nurses is kept at the hospital, and any one wishing to engage a nurse communicates with the matron, stating whether a pupil nurse or a graduate is desired, the condition of the patient, and other details. The charge made for a pupil nurse is ordinarily $10 or $12, and for a graduate from $15 to $21 per week.

The instruction given in a training school for nurses includes courses of lectures by the physicians, in addition to personal instruction by the superintendent and matron, on such subjects as anatomy, physiology, hygiene, diseases (general and contagious), surgery, materia medica, and obstetrics. The matron also gives lessons on practical nursing and ward-work. At the close of each year a written examination takes place in the theoretical and practical work of the course prescribed by the regulations; and an examination, under the direction of the members of the Executive Committee, is held every six months. The management of the training school is in the hands of the matron, who is assisted by an advisory board. The period of training is two years as a rule at the present time, but we are glad to see that in one case at any rate the course has been lengthened to three years. After a month's probation the matron decides as to the propriety of retaining the applicant as a pupil nurse. A pupil nurse may at any time be discharged in case she proves inefficient, and may at any time be suspended or discharged, for negligence or misconduct, by the matron. The prescribed age is from twenty-three to thirty-five years, and every pupil nurse must have received a good common school education, and must be provided with a certificate of good health from her physician. Each pupil nurse binds herself to remain at the school for a term of two years, subject to its rules and regulations, and during that time she is expected to serve in the wards and in the homes of the sick, of whatever condition, to which she may be sent.

The pupils are boarded and lodged in the hospital during the month of probation, and if approved they receive about $7 per month for the first five months, $9 per month for the next six months, and $12 per month for the second year. These payments are given not as compensation for services rendered, for which their education is considered a sufficient recompense, but as a means to provide for the necessities of hospital life. Pupils wear the hospital uniform, the material for which is provided by the superintendent. Every pupil is allowed two weeks' vacation each year. Pupils completing the two years' course satisfactorily receive diplomas signed by the President, the Secretary, and the Superintendent of Nurses.

The Isolation of Contagious Diseases.—Formerly in
England it was the common practice for a community to
make arrangements with the authorities of the general
hospitals to provide a separate block in which cases of
fever and other infectious diseases could be treated.
During the last twenty years legislation has taken place
which casts the duty of providing such accommodation
upon the local Sanitary Authorities all over the country,
and as a consequence general and cottage hospitals in
Great Britain, with a very few exceptions, have discontinued the practice of treating infectious disease.

In the United States, with the exception of the
large centres of population, very little or no accommodation is provided for the isolation of fever, small-
pox, and other contagious cases. This fact has led
to the inclusion in the scheme of cottage hospitals
of separate buildings for the treatment of these
diseases. The managers have felt it their duty to
endeavour to educate the community as to the propriety and benefit of the isolation of infectious cases.
With the exception of smallpox, public opinion did
not recognise the necessity for the strict isolation of
contagious diseases, and the removal of the patients
to separate buildings, not only for their own better
treatment, but to guard the community from risk, was
not acceptable to the feelings of the people. Thanks
to the cottage hospital movement, this feeling has
now altered ; and where no adequate hospital provision has been in existence, the cottage hospitals have
succeeded in inducing the city authorities to appro-

priate the means for the construction of contagious wards, the management of which has been entrusted to the Executive Committee of the Cottage Hospital. The wisdom of this expenditure has been shown not only in the treatment of cases sent in to these wards, but in the security felt by the citizens as they have heard of the difficulties which people in other places have experienced in improvised hospitals for contagious cases when epidemics have appeared. Much remains to be done, however, with a view to overcoming popular reluctance in the removal of the sick from their own homes. Affection for the afflicted ones and ignorance of the danger to others still check what would be the kindest thing to do for all parties.

Maternity Cottages. —At Newton and a few other places a maternity cottage has been provided for the reception of women who cannot obtain at home the comforts and conveniences needed at the critical period of maternity. The conditions of admission include a regulation confining free cases to residents within the boundary of the district served by the hospital. Non-residents are, however, admitted with the consent of the Executive Committee. Where the patients are in a position to pay, residents are charged a fee of $10, and non-residents one of $15 and upwards, according to the nature of the case. This charge includes medical attendance, board and care for two weeks, and for such further time as may be deemed necessary by the medical directors. The arrangements are in charge of the matron, who must receive notice at least one week before the expected date of confinement. Each patient must be previously examined by the attending physician of the maternity cottage. Patients are not expected to present themselves for admission before the day of confinement or until labour has commenced, and, except in complicated cases, they are discharged at the end of two weeks after confinement. Accommodation is provided for the reception of private patients, who are received under the same conditions in private wards attached to the cottage hospital proper. It is estimated that the cost of a maternity cottage need not exceed $6000.

Convalescent Cottages.—The movement is extending in favour of the provision of convalescent cottages for the reception of patients discharged from the cottage hospital before they are physically capable of resuming their daily work, and for the reception of other persons who, though free from any special disease, still need a few weeks' rest and nursing. The patients under treatment consist of two classes—(1) those able to be about, although not cured ; and (2) those suffering from weakness or physical depression, who pay a small sum per week for their board. It is estimated that the cost of erecting a convalescent cottage would be about $5000, and that the furniture would absorb another $1000, making a total cost of $6000.

Hospital Cottages for Children.—An interesting system of cottage hospitals for children has been in operation for about ten years at Baldwinsville, on the

Fitchburg Railroad in Massachusetts. Patients from
all parts of Massachusetts, including very many from
Boston, have been received, and the success attained
was so great that after four years' experience it was
deemed wise to enlarge its sphere and to much extend
the area of its work. Competent surgical and
medical treatment are provided at these hospital
cottages, together with such educational and moral
training as may be best adapted to the limited
capacities of each child.

There are three classes of incurable children for whom
provision is made. First, those rendered invalids for
life from congenital diseases, paralysis, and kindred
ailments; secondly, epileptic children; and, thirdly,
the class known as asylum cases, whose peculiar needs
require separate rooms and especial care. The
hospital cottages for children have now been incorporated, and they provide care, training, and treatment for diseased, maimed, paralytic, feeble-minded,
destitute, and orphan children. The majority of the
patients or their friends pay a small weekly sum, but
there are in addition a number of free patients. No
child is ever refused admission provided there be
room for its reception. The school connected with
these cottages is doing a good work for children otherwise deprived of mental training. It may interest the
reader to learn that each room contains two iron beds,
with wire and hair mattresses, bedding, pillows, etc.,
a bureau with mirror, stained chairs, and pictures. The cost of furnishing such a room is about

$50 or £10. There are single rooms with one bed, the cost of furnishing which is $30 or £6. No children of either sex over 17 years of age are at present received, but it is in contemplation to make provision for a certain number of adult cases. In connection with the hospital cottages for children there is a large farm with a farmhouse, where several of the larger boys have an agreeable home. Altogether, these hospital cottages for children deserve the warmest support and encouragement at the hands of the benevolent. The excellence of the work done is admitted on all sides. The principle of hospital cottages has now won its way to recognition in the United States, and a number of similar hospitals have been established elsewhere. It is interesting to note that by an Act of the Commonwealth of Massachusetts, approved April 9th, 1889, the State Treasurer is authorised to allow and pay to the hospital cottages for children a sum not exceeding $55,000, to be expended for the purchase of land and the erection of buildings suitable for the accommodation of the inmates, subject to the approval of the plans and estimates by the Governor and Council of the State.

Cottage Hospital Construction.—As may well be imagined, the cottage hospital movement in the United States has found little or no difficulty in regard to sites. The sites of most of the existing institutions are of ample proportions, and they seem to be well chosen for the purpose to which they are put. The history of the cottage hospital movement in the United States shows that, as a rule, it has been the practice to hire a suitable house or other building in the first instance, and to adapt it for the reception of patients. The size of the original building has usually been greater in the United States than in this country; and it will be noticed that there is only one cottage hospital on our list which contains so few as seven beds. Where the managers have adapted an existing house or building they have had the wisdom to make as few alterations as possible, and have contented themselves in the majority of cases with the introduction of necessary sanitary improvements and an adequate heating apparatus. The former has frequently included the abolition of an existing cesspool and the construction of a subsidiary sewer, which has been connected with the main drainage system of the township where the cottage hospital is situated. We have had some difficulty in selecting a plan for reproduction and description. The fact is, that although several of the hospitals have published a plan of their institutions in the reports they have issued, very few of them have been drawn to scale, and the diagrams published are therefore of no value for practical purposes. We cannot too earnestly urge upon those concerned the importance of publishing plans to scale in every case where the managers determine to go to the expense of preparing blocks for use in the annual reports.

The difficulty we have had in selecting a plan in any way typical of the kind of buildings used for

cottage hospital purposes in the United States is of
relatively small importance. This is so because a
building which is to be devoted to the care of the
sick, in order to fulfil hygienic conditions, when once
planned upon the most approved principles, is easily
capable of being adapted to the peculiar circumstances
incidental to the climate of a particular country.
Hence the model and other plans of cottage hospitals
and kindred institutions which we publish in Chapters
IX. and X. of this book will no doubt prove
almost if not quite as useful and interesting to residents in the United States as they have already been
found useful to cottage hospital managers in this
country. Those who are interested in the question
of climate as applying to hospital construction should
refer to Vol. IV. of *Hospitals and Asylums of the
World*,[1] where the matter is fully dealt with. That
volume is devoted entirely to hospital construction and kindred subjects, and any one who refers
to it will find a mass of information covering the
whole ground so far as the principles of planning,
the special requirements of hospitals, and all the
technical points are concerned.

We believe it would be accurate to state that of
cottage hospitals in the English meaning there are
few or none in the United States. Of small hospitals
for urban communities there are, however, quite a
number, as will be seen from our list ; and, having

---

[1] *Hospitals and Asylums of the World.* By Henry C. Burdett.
London : The Scientific Press. Limited, 428 Strand, W.C.

regard to the fact that the majority of them have fifty beds or under, they have been classed as cottage hospitals, and their names given in this book.[1] The plan favoured in the United States for small hospitals, owing, no doubt, in a measure to the fact that a suitable and large site is as a rule obtainable, is a small edition of the plan adopted at Hamburg in the New General Hospital at Eppendorf, which is one of the most complete in the world.

The method pursued is to provide that each department of the hospital shall be housed in a separate building. Thus there is a medical building, a surgical building, an administrative building, and one or more wards for non-contagious cases, the whole being connected by corridors. In addition to this, provision is frequently made for a contagious ward, isolated and quite apart from the other buildings, which is approached by a separate entrance, and is entirely cut off from other portions of the hospital, including the grounds. There is sometimes too a separate nurses' home, with accommodation for a fairly large staff, owing to the plan of training nurses at small hospitals in the United States. The laundry building, an ambulance and boiler house, and a mortuary complete the buildings which an ambitious and small community will provide for the accommodation of the sick of all grades within the area of their jurisdiction. No hospital in the United States, small or large, would be complete without a building devoted to

[1] See p. 363.

private wards for the reception of paying patients. These wards are very comfortably fitted up, and no one who has ever had the privilege of being treated in them would ever wish to go elsewhere when ill. The private ward or pay pavilion possesses the additional advantage of affording the managers an opportunity of fully training their nursing staff in all the duties and varied difficulties which beset a nurse in a private house.

In England nurses who have been trained in a great public hospital, and on the completion of their training determine to devote themselves to private nursing, are often placed at a serious disadvantage, because what is perfectly natural and fitting in a public institution may be wholly inconvenient and even improper where the nursing has to be done in a private house. We have often thought that the nurses who are trained in the United States have a considerable pull over British nurses, owing to the advantages the former derive from the experience gained in the private or pay wards. Is it too much to hope that the authorities of English nurse training schools will consider the point here raised, and take steps to move the hospital authorities to institute private wards for the reception of paying patients in connection with every hospital? This step, if universally taken, would be of immense advantage to the public, indirectly owing to the better training which the nurses would thus receive, and directly from the circumstance that a large class of patients, able and willing to pay a remunerative rate for their

24

treatment when in hospital, are at present deprived of the privilege, because English hospital managers have so far declined to recognise the situation by introducing the pay ward for their reception. We are confident much of the hospital abuse in England which at present cries aloud for reform would disappear, if pay wards were established in connection with every general hospital throughout the country.

COMMUNICATIONS BETWEEN THE AUTHOR AND COTTAGE HOSPITAL MANAGERS.

Points upon which the managers of Cottage Hospitals are requested to give special information in any communication they may address to the Author.

In all cases please send a copy of the last report of the cottage hospital under your management, with the bye-laws and the following particulars :—

(1.) The date of its foundation, and the endowment, if any.

(2.) The cost of its erection, and how the amount was raised, *viz.*, by subscription or otherwise. Did the farmers and others n the district assist the scheme with presents of stone, wood, and labour? Was it erected at the cost of one person alone?

(3.) If it is a cottage, altered to meet the requirements of a hospital, what rent do you pay, and what was the cost of the alterations? Could you send me a plan of it— ground plan and elevation?

(4.) What is the extent of the district from which patients are sent, and what is the name of the nearest town having a general hospital?

(5.) Do the patients pay anything for the relief they receive, and what is the rule with regard to pauper cases?

(6.) Do the medical staff receive payment for their services?

(7.) Average income and average expenditure : average number of beds occupied during past three years.

(8.) Number of beds ; number of patients.

(9.) System of nursing. Number of nurses.

If this information can be furnished, together with any other on points which may be considered of importance and interest, a great service will be rendered to the author.

(355)

Form Suggested to Cottage Hospital Managers for general
INCOME AND EXPENDITURE ACCOUNT

INCOME.

£ s. d. £ s. d.

A. ORDINARY

    I. Annual Subscriptions (see page) - -

    II. Donations (see page) - - - -

        Boxes (see page) - - - - -

    III. Hospital Sunday Fund - - -

    IV. Hospital Saturday Fund - -

    V. Workpeople's Contributions (apart from
        Saturday Fund) - - - - -

    VI. Congregational Collections (apart from
        Sunday Fund) - - - - -

    VII. Entertainments - - - -

    VIII. Invested Property—
        Dividends - - - - -
        Income Tax Returned - -
        Interest on Deposit Account -
        Rents (Net) - - - -

    IX. Nursing Institution—
        Private Nurses - - - -
        Nurses' and Probationers' Fees - -

    X. Patients' Payments—
        In-Patients - - - -
        Out-Patients - - - -

    XI. Other Receipts—

        Total Ordinary Income - -

B. EXTRAORDINARY—

    Legacies (see page) - - - - -

CAUTION.—This form in its entirety is too elaborate for cottage hospital purposes. It is given, however, to enable the Secretaries to select those items which apply to their institutions, and so to classify their income and expenditure accurately.

£

OF ACCOUNTS.
consideration, and to secure uniformity of Account.
for the year ending 31st December, 189 .

EXPENDITURE.

		£ s. d.	£ s. d.

A. MAINTENANCE. I. Provisions—
    Meat
    Fish, Poultry, etc.
    Butter, Cheese, etc.
    Eggs
    Milk
    Bread, Flour. etc.
    Grocery
    Vegetables
    Malt Liquors

  II. Surgery and Dispensary—
    Drugs, Chemicals, Disinfectants, etc.
    Dressings, Bandages, etc.
    Instruments and Appliances
    Ice and Mineral Waters
    Wine and Spirits

  III. Domestic—
    Renewal of Furniture
    Bedding and Linen
    Hardware, Crockery, Brushes, etc.
    Washing
    Cleaning and Chandlery
    Water
    Fuel and Lighting
    Uniforms

  IV. Establishment Charges—
    Rates and Taxes
    Insurance
    Garden
    Annual Cleaning
    Repairs (Ordinary)

  V. Rent
  VI. Salaries. Wages, etc.—
    Medical
    Dispensing
    Nursing
    Other Salaries and Wages
    Pensions

  VII. Miscellaneous Expenses—
    Printing, Stationery, Postage, and Advertisements
    Sundries

B. ADMINISTRATION I. Management—
    Official Salaries
    Commission
    Pensions
    Official Printing and Stationery
    Official Postage and Telegrams
    Official Advertisements
    Law Charges
    Interest on Loan
    Auditors' Fee
    Sundries

  II. Finance—
    Appeals
    Festival

    Total Ordinary Expenditure
C. EXTRAORDINARY—I. Repairs
  II. Building Improvements

    Total Extraordinary Expenditure

£

A MODEL CASE BOOK FOR COTTAGE HOSPITALS.

Mr. Thomas Moore's Case Book, as used at Petersfield Cottage Hospital.

No.	Name.	Date of Admission.	Age.	Residence.	Recommending Subscriber.	Payment Weekly. s. d.	By whom Guaranteed.	Date of Payment.	Amount Paid. s. d.	By whom Paid.	Remarks.	Disease.	Result.	Date of Discharge or Death.	Number of Days in Hospital.													Medical Attendant.
															Jan.	Feb.	Mar.	April	May	June	July	Aug.	Sept.	Oct.	Nov.	Dec.	Total.	

ALPHABETICAL LISTS OF COTTAGE HOSPITALS IN (*a*) THE UNITED KINGDOM, AND (*b*) THE UNITED STATES OF AMERICA.

EXPLANATORY NOTE.

In the second edition of this book we published four appendices (D to G), containing (1) an alphabetical list of cottage hospitals, with particulars of their income and expenditure; (2) an alphabetical list of cottage hospitals, showing their sources of income; (3) an alphabetical list of cottage hospitals, showing endowments, cost of erection, and other information; and (4) an alphabetical list of cottage hospitals, showing rent paid, number of beds, patients, salaries of medical officers, rates of patients' payments, etc. We have decided to omit these alphabetical lists from the present book, because, so far as (1), (2), and (3) are concerned, the particulars are already given accurately each year in *Burdett's Hospitals and Charities*, being the Year Book of Philanthropy (London : The Scientific Press, Limited); and experience shows that it is better to have the information up to date. Any one therefore who desires to obtain detailed information concerning any individual cottage hospital will find it on reference to the last annual edition of *Burdett's Hospitals and Charities*. If the information is desired for more than one year it may be obtained by a reference to the back volumes of *Burdett*. The special particulars contained in (4) have not altered much since they were first published in the second edition of *Cottage Hospitals*, which was issued from the press in 1880. We have determined to add these particulars as far as they are material to those formerly given in *Burdett's Hospitals and Charities*, as by this means it will be possible to immediately note all the changes which may take place from time to time, and so to secure the latest and most accurate information for the use of those who are interested in any special point.

(*a*) LIST OF COTTAGE HOSPITALS IN THE UNITED KINGDOM.

Aberchirder Cottage Hospital (Banff).

Aberdare Cottage Hospital (Glamorganshire).

Abergavenny Cottage Hospital (Monmouthshire).

Abingdon Cottage Hospital (Berks).

Alloa Cottage Hospital (Clackmannan).

Andover Cottage Hospital (Hampshire).

Ashburton and Buckfastleigh Cottage Hospital (Devon).

Ashby-de-la-Zouch Cottage Hospital (Leicestershire).

Ashford Cottage Hospital (Kent).

Axminster Cottage Hospital (Devon).

Balfour Hospital, Kirkwall (Orkneys).

Ballymena Cottage Hospital (Antrim).

Ballyshannon (Donegal), Shiel Cottage Hospital.

Bangor Cottage Hospital (Co. Down).

Barnet (Herts), Victoria Cottage Hospital.

Barnsley (Yorks), Beckett Hospital and Dispensary.

Barrow-in-Furness (Lancashire), North Lonsdale Hospital.

Barry Cottage Hospital (Glamorganshire).

Barton-under-Needwood Cottage Hospital (Stafford).

Basingstoke Cottage Hospital (Hants).

Batley and District Cottage Hospital (Yorkshire).

Beccles Hospital (Suffolk).

Beckenham Cottage Hospital (Kent).

Berkeley Hospital (Gloucester).

Betteshanger Cottage Hospital (Kent).

Beverley Dispensary and Hospital (Yorkshire).

Bexley Cottage Hospital (Kent).

Bingley Cottage Hospital (Yorkshire).

Blackheath and Charlton Cottage Hospital (Kent).

Blaenau Festiniog (Merioneth), Oakeley's Hospital.

Blandford Cottage Hospital (Dorset).

Bodmin, East Cornwall Hospital (Cornwall.)

Boscombe Hospital (Hants).

Boston Hospital (Lincolnshire).

Bournemouth (Hants), Royal Victoria Hospital.

Bourton-on-the-Water Cottage Hospital (Gloucestershire).

Brackley Cottage Hospital (Northamptonshire).

Braintree and Bocking Cottage Hospital (Essex).

Brandon, Northwold Cottage Hospital (Suffolk).

Brentwood Cottage Hospital (Essex).

Bridgend Cottage Hospital (Glamorganshire).

Bridlington, Lloyd Cottage Hospital (Yorkshire).

Bridport Dispensary and Cottage Hospital (Dorset).

Bromley Cottage Hospital (Kent).

Bromley (Kent), Phillips Memorial Homeopathic Hospital.

Bromsgrove Cottage Hospital (Worcester).

Bromyard Cottage Hospital (Hereford).

Brotton (Yorkshire), Cleveland Cottage Hospital.

Buckhurst Hill Village Hospital (Essex).

Budleigh Salterton Cottage Hospital (Devon).

Burford Cottage Hospital (Oxon).

Burslem (Staffs), Haywood Hospital.

Capel Village Hospital (Surrey).

Chalfont St. Peter's Cottage Hospital (Bucks).

Charlwood Cottage Hospital (Surrey).

Chelmsford, Essex and Chelmsford Infirmary (Essex).

Chesham Cottage Hospital (Bucks).

Cheshunt Cottage Hospital (Herts).

Chislehurst, Sidcup, and Cray Valley Cottage Hospital (Kent).

Chorley Dispensary and Cottage Hospital.

Chulmleigh Cottage Hospital (Devon).

Cirencester Cottage Hospital (Gloucester).

Clevedon Cottage Hospital (Somerset).

Clun Cottage Hospital (Salop).

Cold Ash (Berks), Children's Cottage Hospital.

Coldstream Cottage Hospital.

Coleraine Cottage Hospital (Londonderry).

Congleton Cottage Hospital (Cheshire).

Cork, Valencia Village Hospital.

Cranleigh Village Hospital (Surrey).

Crewe Memorial Cottage Hospital (Cheshire).

Crewkerne Hospital (Somerset).

Cromer Cottage Hospital (Norfolk).

Croydon General Hospital (Surrey).

Cumnock Cottage Hospital (Ayrshire).

Darlington Cottage Hospital for Children (Durham).

Dartford (Kent), Livingstone Memorial Cottage Hospital.

Dartmouth Cottage Hospital (Devon).

Dawlish Cottage Hospital (Devon).

Devizes Cottage Hospital (Wilts).

Dewsbury and District General Infirmary (Yorkshire).

Dingwall, Ross Memorial Hospital.

Dinorben Cottage Hospital (Anglesey).

Dinorwic Quarry Hospital (Carnarvon).

Ditchingham (Suffolk), All Hallows Country Hospital.

Dorking Cottage Hospital (Surrey).

Dover Hospital (Kent).

Driffield Cottage Hospital (Yorkshire).

Drogheda Memorial Cottage Hospital (Co. Louth).

Dufftown (Banff), Stephen Cottage Hospital.

Dunfermline Cottage Hospital (Fife).

Dunster and Minehead Village Hospital (Somerset).

Ealing Cottage Hospital (Middlesex).

East Grinstead Cottage Hospital (Sussex).

Eastbourne (Sussex), Princess Alice Memorial Hospital.

Eastbourne (Sussex), Victoria Cottage Hospital.

Egham Cottage Hospital (Surrey).

Eltham Cottage Hospital (Kent).

Emsworth and District Jubilee Cottage Hospital (Hants).

Enfield Cottage Hospital (Middlesex).

Epsom and Ewell Cottage Hospital (Surrey).

Erith, Crayford, Belvedere, and Abbeywood Cottage Hospital (Kent).

Evesham Cottage Hospital (Worcester).

Exmouth (Devon), Maud Hospital.

Fairford Cottage Hospital (Gloucester).

Falmouth Cottage Hospital (Cornwall).

Faringdon Cottage Hospital (Berks).

Faversham Cottage Hospital (Kent).

Fleetwood Cottage Hospital (Lancs).

Forgue Cottage Hospital (Aberdeen).

Forres Leanchoil Cottage Hospital (Elgin).

Fort William, N.B., Belford Hospital.

Fowey Cottage Hospital (Cornwall).

Frome Cottage Hospital (Somerset).

Fyvie Cottage Hospital (Aberdeen).

Garston Accident Hospital (Lancs).

Goole Cottage Hospital (Yorkshire).

Gorleston Cottage Hospital (Norfolk).

Grantham Hospital (Lincoln).

Grantown (Elgin), Ian Charles Cottage Hospital.

Guernsey, Victoria Cottage Hospital.

Guisborough Miners' Accident Hospital (Yorkshire).

Halstead Cottage Hospital (Essex).

Hambrook Village Hospital (Gloucester).

Hammerwich and District Cottage Hospital (Staffs).

Harrogate Cottage Hospital (Yorkshire).

Harrow Cottage Hospital (Middlesex).

Hatfield Broad Oak Cottage Hospital (Essex).

Hawick Cottage Hospital (Roxburgh).

Haydock Cottage Hospital (Lancashire).

Hayes Cottage Hospital (Middlesex).

Herne Bay Cottage Hospital (Kent).

High Wycombe and Earl of Beaconsfield Memorial Cottage Hospital (Bucks).

Hillingdon Cottage Hospital (Middlesex).

Hinckley Cottage Hospital (Leicestershire).

Holyhead (Anglesey), Stanley Hospital.

Horsham Cottage Hospital (Sussex).

Hounslow Hospital (Middlesex).

Huntly Jubilee Cottage Hospital (Aberdeen).

Ilfracombe, Tyrell Cottage Hospital (Devon).

Ilkeston Cottage Hospital (Derby).

Iver, Langley, and Denham Cottage Hospital (Bucks).

Jarrow-on-Tyne Memorial Hospital (Durham).

Johnstone Cottage Hospital (Renfrewshire).

Keighley and District Hospital (Yorkshire).

Keith (Banff), Turner Memorial Hospital.

Kendal Memorial Hospital (Westmoreland).

Keswick Infirmary (Cumberland).

Kettering Hospital (Northamptonshire), in course of construction.

Kington (Hereford), Victoria Cottage Hospital.

Kirkcaldy Cottage Hospital (Fife).

Lanark, Lochart Hospital.

Ledbury Cottage Hospital (Hereford).

Leek (Staffs), Memorial Cottage Hospital.

Leyton, Walthamstow, and Wanstead Children's General Hospital (Essex).

Liskeard (Cornwall), Passmore Edwards' Cottage Hospital, in course of construction.

Littlehampton (Sussex), Cottage Hospital.

Llandrindod Wells(Radnorshire),Cottage Hospital and Convalescent Home.

Llandudno (Carnarvonshire), Sarah Nicol Memorial Cottage Hospital.

Llangollen Cottage Hospital (Denbigh).

Longton Cottage Hospital (Staffs).

Loughton (Essex), Oriolet Cottage Hospital.

Louth Dispensary and Hospital (Lincs).

Ludlow Cottage Hospital (Salop).

Luton, Bute Cottage Hospital (Bedford).

Lydney Cottage Hospital (Gloucester).

Lyme Regis Cottage Hospital (Dorset).

Lynton District Cottage Hospital (Devon).

Lytham Cottage Hospital and Convalescent Home (Lancs).

Maidenhead Cottage Hospital (Berks).

Malton Cottage Hospital (Yorkshire), proposed.

Malvern Rural Hospital (Worcester).

Mansfield and Mansfield-Woodhouse Dispensary and Hospital (Notts).

Market Rasen Cottage Hospital (Lincs).

Melksham Cottage Hospital (Wilts).

Mexborough (Yorkshire), Montagu Cottage Hospital.

Middlesbrough (Yorkshire), North Ormesby Cottage Hospital.

Mildenhall Cottage Hospital (Suffolk).

Mildmay Park, London, N., Memorial Cottage Hospital.

Milton Abbas Cottage Hospital (Dorset).

Mirfield Memorial Hospital (Yorks).

Mold Cottage Hospital (Flint).

Monkwearmouth and Southwick Hospital (Durham).

Moreton-in-Marsh Cottage Hospital (Gloucester).

Mount Sorrel Cottage Hospital (Leicester).

Mull Cottage Hospital (Argyle).

Newbury District Hospital (Berks).

Newick Cottage Hospital (Sussex).

Newmarket (Cambridge), Rous Memorial Hospital.

Newton Abbot (Devon), Newton Cottage Hospital.

Northallerton Cottage Hospital (Yorks).

North Shields (Northumberland), Tynemouth Victoria Jubilee Infirmary.

Northwich (Cheshire), Victoria Infirmary.

Norwood Cottage Hospital.

Nuneaton and District Cottage Hospital (Warwickshire).

Ormskirk Cottage Hospital (Lancs).

Oswestry and Ellesmere Cottage Hospital (Salop).

Ottery St. Mary |District Cottage Hospital (Devon).

Paignton Cottage Hospital (Devon).

Paulton Memorial Cottage Hospital (Somerset).

Petersfield Cottage Hospital (Hampshire).

Petworth Cottage Hospital (Sussex).

Plaistow (Essex), St. Mary's Cottage Hospital.

Poole (Dorset), Cornelia Hospital.

Port Talbot Cottage Hospital (Glamorganshire).

Potter's Bar Cottage Hospital (Middlesex).

Redditch, Smallwood Hospital (Worcester).

Redhill (Surrey), Reigate and Redhill Cottage Hospital.

Redruth (Cornwall), West Cornwall Miners' Hospital.

Retford General Dispensary and Cottage Hospital (Notts).

Ripon Dispensary and Cottage Hospital (Yorks).

Romford (Essex), Victoria Cottage Hospital

Rotherham Hospital (Yorkshire).

Royston Cottage Hospital (Cambridge).

Ruabon Accident and Cottage Hospital (Denbigh).

Rugeley District Hospital (Staffs).

St. Albans Hospital (Herts).

St. Andrews (Fife), Memorial Cottage Hospital.

St. Helen's Cottage Hospital (Lancashire).

St. Leonards (Sussex), Buchanan Cottage Hospital.

Saffron Walden Hospital (Essex).

Savernake Hospital (Wilts).

Seacombe Cottage Hospital (Cheshire).

Shaftesbury (Dorset), Westminster Memorial Cottage Hospital.

Shedfield Cottage Hospital (Hampshire).

Shepton Mallet District Hospital (Somerset).

Sherborne, Yeatman Hospital (Dorset).

Shipley (Yorkshire), Sir Titus Salt's Hospital.

Sidcup Cottage Hospital (Kent).

Sidmouth Cottage Hospital (Devon).

Skye, Mackinnon Memorial Hospital (proposed).

Southampton, St. Mary's Cottage Hospital, Northam (Hampshire).

Southend (Essex), Victoria Hospital.

Spalding (Lincs), Johnson Hospital.

Speen Cottage Hospital (Berks), now a convalescent home.

Stockton and Thornaby Hospital (Durham).

Stony Stratford Cottage Hospital (Bucks).

Stourbridge (Worcester), Corbett Hospital.

Stratton Cottage Hospital (Cornwall).

Sudbury (Suffolk), St. Leonard's Hospital.

Surbiton Cottage Hospital (Surrey).
Swaffham (Norfolk), Victoria Jubilee Hospital.
Swanage Cottage Hospital (Dorset).
Swindon (Wilts), Victoria Hospital.
Swindon (Wilts), Great Western Railway Medical Fund Society's Accident Hospital.
Tamworth Cottage Hospital (Staffs).
Tarves, Haddo House Cottage Hospital.
Tavistock Cottage Hospital (Devon).
Teddington and Hampton Wick Cottage Hospital (Middlesex).
Tenbury (Salop), St. Mary's Cottage Hospital.
Tenby Cottage Hospital (Pembroke).
Tetbury Cottage Hospital (Gloucester).
Tewkesbury Rural Hospital (Gloucester).
Thames Ditton Cottage Hospital (Surrey)
Thirsk (Yorkshire), Lambert Memorial Hospital.
Totnes Cottage Hospital (Devon).
Trowbridge Cottage Hospital (Wilts).
Twickenham (Middlesex), St. John's Hospital.
Ulverston and District Cottage Hospital (Lancashire).
Walker Hospital (Northumberland).
Wallasey Cottage Hospital (Cheshire).
Wallingford (Berks), Morrell Memorial Cottage Hospital.
Walsall and District Hospital (Staffordshire).
Wantage Cottage Hospital (Berks).

Warminster Cottage Hospital (Wiltshire).
Warwick Dispensary and Cottage Hospital (Warwick).
Watford District Cottage Hospital (Herts).
Watlington Cottage Hospital (Oxford).
Wellington District Cottage Hospital (Somerset).
Wells Cottage Hospital (Somerset).
Weston-super-Mare Hospital (Somerset).
Whitchurch Cottage Hospital (Salop).
Willesden Cottage Hospital.
Willingham-by-Stow, Gainsborough (Lincs), Reynard Cottage Hospital.
Wimbledon Cottage Hospital (Surrey).
Winchcomb Cottage Hospital (Gloucestershire).
Wirksworth Cottage Hospital (Derbyshire).
Wisbeach (Cambridge), North Cambridgeshire Hospital.
Woodbridge (Suffolk), Seckford Hospital.
Wood Green (Middlesex), Passmore Edwards' Cottage Hospital.
Woolwich and Plumstead Cottage Hospital (Kent).
Wotton-under-Edge Cottage Hospital (Gloucester).
Workington Infirmary (Cumberland).
Wrington Vale Cottage Hospital (Somerset).
Yeovil District Hospital (Somerset).
Yoxall Cottage Hospital (Staffs).

(b) LIST OF COTTAGE HOSPITALS IN THE UNITED STATES OF AMERICA.

List of Hospitals in the States of Massachusetts, New Hampshire, Rhode Island, Maine, Vermont, and Connecticut.

PLACE.	NAME.	Opened.	No. of Beds.	Daily average No. occupied in 1894.
State of Massachusetts—				
Beverley - - -	Beverley Hospital - -	1893	—	9
Boston - -	Vincent Memorial Hospital -	1890	10	—
Brockton - -	Brockton Hospital - - -	—	—	—
Cambridge - -	Cambridge Hospital - - -	1886	38*	27
Clinton - -	Clinton Hospital - - -	1889	13*	—
Everett - -	Everett Hospital - -	—	—	—
Exeter - -	Exeter Cottage Hospital - -	Proposed.		
Fall River - -	Fall River Hospital - -	1888	13*	—
Fitchburg - -	Fitchburg Hospital - - -	—	—	—
Haverhill - -	City Hospital - - -	1888	17*	—
Holyoke - -	City Hospital - - -	1891	23*	—
Lawrence - -	Lawrence General Hospital -	1876	30	—
Lowell - -	Lowell Hospital - - -	1839	52*	23
Lynn - -	Lynn Hospital - - -	1881	40	28
Malden - -	Malden Hospital - - -	1892	38	16

* Estimated.

PLACE.	NAME.	Opened.	No. of Beds.	Daily average No. occupied in 1894.
Melrose - - -	Melrose Hospital - - -	1893	—	—
Natick - - -	Leonard Morse Hospital - -	Proposed.		
New Bedford - -	St. Luke's Hospital - -	1884	20*	—
Newburyport - -	Anna Jaques Hospital - -	1884	13*	10
Newton - - -	Newton Hospital - -	1886	75*	35
North Adams - -	North Adams Hospital - -	1883	—	—
Pittsfield - - -	House of Mercy - - - -	1874	50	30
Quincy - - -	Quincy City Hospital - -	1889	25	13
Salem - - -	Salem Hospital - - -	1873	20	—
Somerville - - -	Somerville Hospital - -	1893	24*	14
Springfield - - -	Springfield Hospital - -	1870	12	—
Taunton - - -	Morton Hospital - -	1888	20	—
Waltham - - -	Waltham Hospital - -	1885	20	17
Worcester - - -	{ City Hospital - - -	1871	65	61
	{ Memorial Hospital - -	1871	40*	22
State of New Hampshire—				
Claremont - - -	Claremont Cottage Hospital -	1893	8	—
Concord - - -	Margaret Pillsbury General Hos.	1884	40	10
Hanover - - -	Mary Hitchcock Hospital - -	—	—	—
Manchester - -	{ Sacred Heart Hospital -	1892	30†	—
	{ Elliott Hospital - - -	1889	18	10
Portsmouth - -	Portsmouth Cottage Hospital -	1885	13	—
State of Rhode Island—				
Newport - - -	Newport Hospital - - -	1873	40	27
Woonsocket - -	Woonsocket Hospital - -	1873	18*	11
State of Connecticut—				
Norwich - - -	William W. Backus - - -	1893	63*	15
Stamford - - -	St. John's Hospital and Home -	1887	8*	6
Waterbury - - -	Waterbury Hospital - - -	1889	42*	—
State of Maine—				
Bangor - - -	Bangor General Hospital - -	1892	23*	—
Lewiston - - -	Central Maine General Hospital	1888	30	20
State of Vermont—				
Burlington - - -	Mary Fletcher Hospital - -	1876	52	51
St. Albans - - -	St. Albans Hospital - - -	Proposed.		

Estimated.　　　　　　　　† Increase to 70 in 1895.

A SET OF RULES AND REGULATIONS FOR A COTTAGE HOSPITAL, CODIFIED FROM THE RULES OF EXISTING COTTAGE HOSPITALS.

(Also see Rules of Harrow Cottage Hospital, p. 211.)

1. This hospital shall be called " The ——— Cottage Hospital."

Meetings of Subscribers.

2. An annual general meeting of subscribers to the institution shall be held in *January*, to receive the report and accounts and to transact the general business of the hospital.

3. Fourteen days' clear notice of such meeting shall be given to every subscriber of *10s.* and upwards. Notice shall also be given in one or more local papers. The omission of any individual notice shall not invalidate the meeting.

4. Every subscription shall become due on the *1st of January* in each year, and be considered annual until notice in writing to withdraw it shall have been given to the Secretary.

5. No one whose subscription shall be unpaid shall be entitled to vote in general meeting.

6. Any clergyman or minister making a congregational collection for the hospital shall be considered a subscribing member during the current year.

7. An annual report of the general working of the hospital and a balance sheet of receipts and expenditure shall be published by the Committee and forwarded to every person entitled to vote at the annual meeting.

8. The property of the hospital shall be vested in Trustees chosen by the subscribers in general meeting.

9. The hospital shall be under the direction of a President, *three* Vice-Presidents, a Committee of Management, and a Ladies' Committee.

10. All appointments to vacancies in the offices of Presi-

(365)

dent, Vice-Presidents, Honorary Medical Officers, Honorary Treasurer, Honorary Secretary, or Auditors, shall be vested in the subscribers, and shall be filled up by them at the annual general meeting, or at a special meeting duly convened.

11. Special meetings of the subscribers may be called by the Honorary Secretary on his own responsibility, or at the request of the President, or on a requisition by any *three* members of the Committee of Management, or of *five* subscribers of *10s.* and upwards, sent to the Honorary Secretary: fourteen days' clear notice of such meeting to be given to the subscribers qualified to be present at the annual general meeting, such notice to specify all the business to be brought forward.

Committee of Management.

12. The Committee of Management shall consist of *ten* members elected from annual subscribers of *10s.* and upwards at the annual general meeting in each year, together with the President, Vice-Presidents, Honorary Medical Officers, the Honorary Treasurer, and the Honorary Secretary *ex officio*.

13. The Committee of Management—*three* of whom shall form a quorum—shall meet at the hospital on the *first Tuesday* of every month, to transact the business of the hospital.

Visitors.

14. Two members of the Committee of Management, who shall be appointed at the monthly meeting of that body, shall act as visitors for each month, and shall attend at the hospital every *Thursday*, at *eleven* o'clock.

15. The visitors shall admit fit patients on the recommendation of the medical officers. They shall also inspect the hospital at uncertain times and examine the monthly accounts.

Medical Staff.

16. The Medical Staff shall consist of all duly qualified practitioners resident within the district, who are willing to give their services.

17. Every medical officer of the institution shall have the right to continue the treatment of any patient under his care

at the time of admission. For all other cases the medical officers shall attend at the hospital for a week in rotation and each in succession, the order of attendance to be fixed by themselves.

Matron and Nurses.

18. The appointment of matron and nurses shall be vested in and held during the pleasure of the Committee of Management. Thsee officers shall reside in the hospital, and shall receive such salaries as the Committee of Management may from time to time determine.

Patients.

19. The hospital shall be open to the poor in the following parishes :

preference being given to those not receiving parish relief.

20. Patients shall be required to pay towards their maintenance a weekly sum, the amount of which is to be fixed (according to each patient's circumstances) by the Committee of Management at the first meeting after the patient's admission. Payment may be remitted if the Committee are satisfied that the patient is unable to pay.

21. The following cases are inadmissible : Cases of mental disorder, infectious cases, incurable cases, and maternity cases.

22. Admission to the hospital shall be by means of a letter of introduction[1] from a subscriber of 10s. and upwards, and accompanied by a medical certificate stating the nature of the disease and of how long standing. A form of guarantee to be approved by the Committee of Management shall be printed upon the letter of introduction, and shall be signed by a responsible person at the time of the admission of the patient.

23. Letters of introduction, together with the medical certificate, must be sent to the Committee of Management on *Tuesday* morning by *ten* o'clock, and the applicants must not come to the hospital until they hear from the Honorary Secretary or the matron that they can be received.

24. Cases of urgency and accident will at once be admitted on application at any time.

[1] For a specimen of letter of introduction see p. 55.

25. No patient shall leave the hospital without the permission of the physician or surgeon under whose care he may have been placed, nor enter any ward but his own without leave.

26. Patients who offend against any of the rules of the house or behave themselves irregularly or indecently shall be expelled, and no person after expulsion shall be re-admitted.

27. Any patient may upon his request to the matron be visited by the clergyman of the parish from which he has come or by the minister of the denomination to which he belongs.

28. The relatives and friends of the patients shall be permitted to visit them at such times and subject to such regulations as the Committee may from time to time determine.

29. Any patient who at the time of his admission shall be under the treatment of a medical man not connected with the hospital may be placed under his charge if the patient so requests, with the knowledge of the medical officers of the hospital.

Remunerative Paying Patients.

30. Any suitable patient desirous of having the comforts and nursing of the hospital may, on the certificate of one of the medical officers, and subject to there being room, have the privilege of being admitted on payment in advance of not less than *20s.* per week in the ordinary wards, or *30s.* per week in a private ward.

Ladies' Committee.

31. The domestic arrangements shall be under the supervision of a Ladies' Committee, consisting of *seven* ladies, to be elected at the annual general meeting.

32. The duties of the Ladies' Committee shall be: —

(a) To superintend the domestic arrangements of the hospital.

(b) To give out the stores and make recommendations as to the purchase of the same.

(*c*) To examine the household accounts, and to present them at each monthly meeting of the Committee of Management for allowance and confirmation.

(*d*) To advise the Committee of Management on all points relating to the comfort of the patients.

(*e*) To read to and instruct such patients as may be in a fit state of health.

33. The Ladies' Committee shall meet not less than once a fortnight, shall keep minutes of all their proceedings, and report to each monthly meeting of the Committee of Management.

34. The Committee of Management are empowered to make bye-laws for the better carrying out of the details connected with the working of the hospital.

Alteration of Rules.

35. The foregoing rules shall not be altered nor shall any new rule be adopted except at the annual general meeting or at a special general meeting duly convened, at least fourteen days' previous notice being given of the proposed alteration or addition.

INDEX.

Abyssinian tube wells, 177.
Access to mortuaries, 146.
Accident cases, treatment of, on admission, 95.
Accommodation, increase in hospital, 135.
Accounts, model form of annual statement, 360.
— uniformity in, 30.
Accumulation of endowment fund, 40.
Administration of alcohol, 94.
— of enemata, 98.
— of medicines, 91, 100.
Admission of enteric fever cases, 150, 193.
— of friends, 106.
— of paying cases, 122.
Agricultural districts, size of cottage hospitals for, 149, 156.
Air space of wards, 166.
Alcohol, administration of, 94.
All Hallows' Country Hospital, 216.
Alton, letter of recommendation at, 55.
Ambulance, 321.
America, Hospital Sunday in, 43.
American Cottage Hospitals, their inception, 327; state of affairs in New England States, 328; state of affairs elsewhere, 329; baneful influence of politics on management of municipal hospitals, 330; efficiency of the voluntary as compared with the rate-supported hospital, 331; tendency to increase free beds, 331; condition of the sick in towns, 332; in the country, 332; the constitution of American hospitals, 333; audit committee, 333; finance committee, 334; executive committee, 334; trustees, 334; visitors, 334; religious ministrations, 334; operations on private patients, 335; finance, 335; history of the Morton Hospital, Taunton, Mass., 335; ladies' associations, 336; patients' payments, 337, 338; derivation of income, 337; average weekly cost per in-patient, 337; endowment of beds, 338; income from nurses, 338; medical staff, 339; training school for nurses, 340, 343; matron, 341; contagious diseases, 345; maternity

cottages, 345; convalescent cottages, 347; hospital cottages for children, 347; construction, 349; list of cottage hospitals in the United States, 363.
Amputations at Newcastle Royal Infirmary, 20.
— Dr. Page's statistics, 20.
— in cottage hospitals, 21.
— Schede's statistics, 20.
Annual Report, what it should contain. 21, 22, 23, 24, 25, 30, 55, 74.
Annual subscribers *versus* donors, 36, 37.
Annual subscriptions the secret of the success of cottage hospitals, 36.
Annual subscriptions, amount raised in cottage hospitals, compared with Metropolitan general hospitals, 38.
Arguments against cottage hospitals, 33.
Arm-slings, 316.
Arnott's valves, 170.
Arrangement of wards, 154.
Ashburton and Buckfastleigh Cottage Hospital, consumption of alcohol. 45; description and history of, 220.
Ash-bins, 203.
Ash-closets, 183.
Ashford Cottage Hospital, 251.
Aspect of hospitals, 154.
Assessment of patients' payments, 45.
Assistant nurses, 71.
Automatic sewage meter, 199.
Average cost per bed in general and in cottage hospitals, 32.
Average number of cottage hospitals established yearly, 16, 17.

Bags for ward purposes, 324.
Barnett, Rev. Canon, and boarding out children in farm-houses, 138.
Bath, movable, 314.
Bath-room, 156.
Batley Cottage Hospital, consumption of alcohol at, 95.
Batthyany, Madame, scheme of convalescent cottages, 137.
Beccles Cottage Hospital, 85, 114, 258.
Bedford Institute convalescent cottage system, 133.
Bed, allowance of linen per, 305.
— pans, 325.

(371)

Bed rests, 315.
— sores, 89.
Beds, 304.
— endowment of, in America, 338.
— for paying patients, 122.
— headings for, 311.
-- making, 86.
— number to be stated, 22.
Beer and wine cellar, 163.
Beer, wine, and spirits in cottage hospitals, 95.
Belfast Royal Infirmary, cost per bed at, 32; sources of income at, 35.
Birmingham General Hospital, cost per bed at, 32; sources of income at, 35.
Black, Miss, and chronic hospitals, 240.
Blankets, 304.
Blisters, application of, 98.
Boarding of patients by nurse or matron, 111.
Boarding out children in farm-houses, 136, 138.
Bookshelves, 320.
Boston Cottage Hospital, 184; description and history of, 231; furnishing wards at, 114; nursing at, 73.
Bourton-on-the-Water Cottage Hospital, history of, 299; description of, 257; cost of alcohol at, 95.
Brackets for medicine bottles, 311.
Braintree and Bocking Cottage Hospital, history and description of, 239.
Bread poultices, 98.
Brixham Cottage Hospital, 271.
Bromley Cottage Hospital, history of cases as given at, 25; small wards at, 158.
Bromsgrove Cottage Hospital, cost of alcohol at, 95.
Budleigh Cottage Hospital, cost of alcohol at, 95.
Buildings, old compared with new, 3.
Burdett's Straight and other Model Cottage Hospitals, 290.
Burns, dressing of, 103.

Calorigen stoves, 172.
Cases admitted by house surgeon, 67.
Cases, particulars which should be given in the Annual Reports, 22, 23, 24, 356, 358.
Cases not admissible, 55, 111, 114.
Castor oil, administration of, 91.
Catheter, passage of, 98.
Cause of death should always be stated, 21.
Ceiling to be lime-washed, 162.
Cellar for beer and wine, 163.
Cesspits, 157, 183.
Cesspools for sink-water, 202.
Chalfont St. Peter Cottage Hospital, cost of alcohol at, 95.
Chaplain, 108.
Charing Cross Hospital, cost of alcohol at, 95.
Cheshunt Cottage Hospital, 260.
Chloralum as a disinfectant, 187.

Chronic hospitals, 240.
— cases not admissible, 55, 111.
Church and chapel collections, 42.
Cinder sifter, 203.
Cistern for rain-water, 179.
Civility of nurses, 79.
Cleanliness in cottage hospitals, 82, 117.
Clinical thermometer, use of, 98.
Closets, 32, 156.
Closure of cottage hospitals, 11.
Clothes of patients, 88, 314.
Coal calorigen stove, 173.
Cod-liver oil, administration of, 91.
Collections in churches and chapels, 42.
Construction, bad arrangement of wards 156.
— beer and wine cellar, 163.
— ceilings, 162.
— ends to be kept in view, 289.
— floors, 162.
— general arrangement for hospital with ten beds, 156.
-- general arrangement for hospital with twenty beds and upwards, 160.
— hot water supply, 163.
— laundry, 165.
— matron's sitting-room, 164.
— one or two stories, 159.
— of cottage hospitals in America, 349.
— operating room, 164.
-- scullery, 164.
— the pavilion plan the most useful, 159; but the most expensive for small hospitals, 289.
— ventilation, 166.
— walls, 161.
— warming, 166.
— water supply, 176.
Consultants and country practitioners, 61.
Consultation by medical staff on outside cases, 60.
Convalescent cottages, 133, 136, 347.
— homes, need for, 132, 135.
— homes and cottage hospitals may be under the same management, 132.
— hospital at Milford, Staffs, 261.
— institution, model plan of, 282.
— patients, day room for, 163.
Convalescents should not be treated at cottage hospitals, 130.
Co-operation of clergy, 42.
— of cottage with general hospitals, 19, 32, 66.
Cost of building cottage hospitals, 207, 215, 216, 226, 238, 246, 249, 251, 252, 257, 258, 260, 262, 264, 266, 276, 277, 281, 282, 283, 294, 296, 297.
Cost of building fever hospitals, 285, 288.
— fever ward at Stamford, 257.
— mortuary, 147, 284.
— pavilion hospitals, 264.
— per bed of hospitals, 32.
Cottage hospitals and midwifery cases, 28.

Cottage hospitals and patients' payments, 44.
— and training of nurses, 70, 340, 343.
— as feeders to general hospitals, 66.
— annual subscriptions the secret of the success of, 36.
— an advantage to the country practitioner, 7, 59, 62, 65, 110.
advantages from the public point of view, 132.
as a centre for distributing soup in severe winters, 116.
— as convalescent homes, 130, 261,
- average cost per bed, 32. 264.
better without an out-patient department, 115.
better without dispensaries, 116.
— furnishing of, 113.
- general arrangement for ten beds, 156,
general arrangement for twenty beds and upwards, 158.
in the United States (see American).
model plans for, 278.
mortality in, compared with mortality in general hospitals, 20.
need for mortuary accommodation, 143.
number established each year since 1855, 17.
nursing at, 68.
— provision for remunerative patients in, 122.
— selection of site, 153.
— size of, 149.
— surgical treatment in, 2, 60.
— the excellent results effected by their establishment, 6, 33, 60, 62.
Cottages for children at Baldwinsville, Mass., 347.
Cottages for contagious diseases in the United States, 345.
Cottages for maternity cases in the United States, 346.
Counterpanes, 306.
Counties without cottage hospitals, 8.
Country holidays for children, 136.
— patients in town hospitals, 66.
— practitioners and London consultants, 61.
— practitioners and the cottage hospital, 7, 59, 62, 65.
County infirmaries and cottage hospitals, 33, 57, 66.
Courtauld, the late Mrs., and Braintree Cottage Hospital, 239.
Cranleigh Village Hospital, 4, 81.
— description of, 206.
— good ventilation at, 81.
Crease's carbolite filters, 330.
Crewkerne, out-patients at, 115.
— rules as to medical staff at, 68.
Crossman, Mr. Edward, on professional intercourse, 65.
Cupboards for wards, 312.

Damp walls, 157.
Dartford, Livingstone Cottage Hospital, 272.
Day-room for convalescents, 163.
Death, cause of, should always be stated, 21.
Deep well water, 178.
Denton's sewage tank for villages, 193, 196, 199.
Diet, 83.
— at Ashburton, 222.
— table, 84.
Diets, distribution from kitchen of, 163.
Dinners for out-patients, 226.
Dirty habits of lower classes, 86.
Disinfecting at a given time, 117.
— closet, 88.
Disinfection of linen, 88, 165.
Dispensary requisites. 303.
Disposal of excreta, 183.
-- of slops, 188, 195, 199, 200, 204.
Ditchingham Cottage Hospital, description and history of, 216.
Donations, 42.
Draw sheets, 87.
Drainage, 188.
— of Grantham Hospital, 244.
— the evils of the " modern systems of." 4.
-- sub-irrigation, 198.
— tanks for flushing, 196.
— laying of, 189, 198.
— testing, 189.
Drains, flushing, 196.
— what to look for in adapting an old house, 203.
Dressing burns, 103.
Dry earth system for urinals, 195.
Drying stoves for earth closets, 186.
Duncan, the late Dr. Matthews, on mortality in child-bed hospitals, 26.
Duties of medical director, 64, 67.
— of ladies' committee, 77.

Ear, foreign bodies in, 102.
Early literature on cottage hospitals, 2.
Earth-closet system, 183.
— movable, 184.
Eassie, the late Mr. W., on ventilation, 167.
East Grinstead Hospital, closure of, 11.
East Rudham Hospital, closure of, 11.
Eastwick, sewage disposal at, 200.
Easy chairs, 320.
Edinburgh Royal Infirmary, cost per bed at, 52; sources of income, 35.
Egham Cottage Hospital, cost of alcohol at, 95.
Emplastrum elemi, 90.
Endowment fund, 40.
Enemata, administration of, 98.
Enteric fever in cottage hospitals, 150.
— cases, treatment of stools, 194.
Epping Convalescent Cottage, 133.
Erection of hospitals, cost of (see Cost).
Excreta, disposal of, 183, 194.
Expenditure at cottage hospitals, 31, 33.

Fæces, disposal of, 183.
Failures of cottage hospitals, 11.
— of occasional nursing system, 69.
Fainting, treatment of, 101.
Falmouth Cottage Hospital, 270.
Family prayers, 84.
Farmers and earth-closets, 187.
Feeding cups, 325.
Fees for medical staff, 46, 50, 122.
Fever, enteric, cases in cottage hospitals, 159, 193.
Fever hospital, model plan of, 284, 285.
— hospitals, 120.
— hospitals, Local Government Board model plans, 285.
— hospitals in the United States, 345.
wards, Stamford, description of, 252.
Field's sewage tank for villages, 193, 196, 199.
Filtration of water before use, 178, 320.
Financial success of cottage hospitals, 33.
Financial success of Eastwick sewerage scheme, 200.
Fireplaces for small cottage hospitals, 169.
First steps to be taken when starting a cottage hospital, 15.
Fits, treatment of, 102.
Floors, 162.
Floor space per bed, 166.
Flowers in wards, 82.
Flushing of drains, 196.
— of urinals, 195.
Fomentations, 100.
Food, administration of, to semi-unconscious patient, 100.
Food, distribution of, from kitchen, 163, uneaten, 88.
Foot rests, 320.
Foreign bodies in ear, etc., 102.
Forres Leanchoil Hospital, 262.
Foundation of a new hospital, 15.
Founder, the, of cottage hospitals, 5.
Fowey Cottage Hospital, boarding system formerly in existence at, 112.
Free medical relief versus patients' payments, 46, 56.
Free medical relief in the United States, 331.
Free system, advantages of, 52.
Friends in cottage hospitals, 106.
— Society of, convalescent cottages, 134.
Frost-bite, treatment of, 102.
Funded property, 40.
Furnishing a cottage hospital, 113.

Galton's stove, 169.
Gamgee, the late Mr., on preservation of ice, 92.
Gas calorigen stoves, 174.
General hospital, model plan of, 282.
— hospitals and cottage hospitals, 19, 32.
— hospitals and patients' payments, 43, 45, 125.

General hospitals, income of, 32.
— — average cost per bed, 32.
— — and annual subscribers, 38.
— — cost of alcohol in, compared with cottage hospitals, 95.
George's calorigens, 173.
German Hospital, cost of alcohol at, 95.
Glasgow Royal Infirmary, cost per bed at, 32; sources of income, 35.
Grantham Cottage Hospital, 95, 243. mortuary at, 143.
Grates, Sir Douglas Galton's stove, 169.
— Pridgin-Teale grate, 172.
— Thermoson grate, 170.
Great Bookham Hospital, closure of, 13.
Great Northern Central Hospital, cost of alcohol at, 95.
Greenhouse, usefulness of, 186.
Guardians, payment for paupers, 31, 51.
Guy's Hospital, cost of alcohol at, 95.

Hæmorrhage, 86, 103.
Hair mattresses, 303.
Hambrook Village Hospital, cost of alcohol at, 95; convalescents at, 131.
Hammerwick Cottage Hospital, cost of alcohol at, 95.
Hammock for surgical cases, 317.
Harris, Rev. F. H., on cottage hospitals, 3.
Harrogate Cottage Hospital, history and description of, 215.
Harrow Cottage Hospital, summary of rules, 211; rules for ladies' committee, 77.
Head, injuries of, 104.
Headings for beds, 311.
Health resorts and infectious diseases, 120.
High Wycombe Cottage Hospital, 258.
Hinckley Cottage Hospital, 9.
Holy Rood, Sisters of the, 5, 132.
Hospital accommodation, increase of, 135.
— cottages for children at Baldwinsville, Mass., 347.
— cottages for women, the, 125.
— kitchen, 115.
— Sunday, 42.
Hospitals for infectious diseases, need for, 120.
Hot water plates, 314.
— — supply, 163, 182.
— — tins, 325.
House slops, disposal of, 153.
— — surgeon, appointment of, 62.
— — duties of, 64, 67.
How to found a cottage hospital, 15.

Ice, preservation of, 92.
Imperfect details of nursing, 14.
Importance of regular food and medicine, 83.
— of ventilation, 81.
Improvidence caused by free medical relief, 46.
Inadmissible cases, 55, 111, 114.

Income, 33.
   of United States Cottage Hospitals, 337.
   — and expenditure of cottage hospitals compared, 33.
   — sources of, 35.
Infectious diseases, need of hospitals for, 120.
Injuries to head, 104.
In-patients, statement to be given in Annual Report, 23, 24.
   rules at Harrow, 212.
   — payments at Ashburton, 221.
   —   — Harrow, 212.
   —   — Leek, 218.
   —   — Lynton, 216.
   —   — Walker, 219.
Inspection of closets, lavatories, etc., 82.
Instruments, 299.
Investments of cottage hospitals, 40, 44.
   — large, deprecated, 40, 48.
   —   — the cause of bad management, 41.
Iron bedsteads, 304.
Isolation of hospitals, 120.
   — hospital, model pavilion plan, 284, 285, 292.
   hospital, Local Government Board model plans, 285.
   — hospitals in the United States, 345.
Iver and Langley Cottage Hospital, cost of alcohol at, 95.

Jubilee cottage hospitals, 17.

Keswick, Mary Hewetson Cottage Hospital, 9.
King's College Hospital, cost of alcohol at, 95.
King's Sutton Hospital, closure of, 13.
Kitchen at cottage hospitals, 115, 156, 291.
   — gardens, 185, 188, 193.
   — refuse for pigs, 203.

Ladies' associations in the United States, 336.
Ladies' committee, 69, 75, 221.
   —   at Harrow, 77.
Lady probationers at cottage hospitals, 73.
Lady superintendent over nursing, 69.
Lavatories, inspection of, 82.
Laundry, 165.
Leeds General Infirmary, cost per bed at, 32; sources of income at, 35.
Leeches, application of, 98.
Leek Memorial Cottage Hospital, description and history of, 217; mortuary at, 143.
Legacies received by cottage hospitals, 33.
Letters of recommendation, 52, 110, 114.
   —   — model, 55.
Limitation of reserve fund, 41.
Linen, 304.
   — disinfection of, 88, 165.

Linseed poultices, 99.
Lint, 308.
Literature, early, on cottage hospitals, 2.
Livingstone Cottage Hospital, Dartford, 272.
Local Government Board, model mortuaries, 145; model isolation hospitals, 285.
Local Sanitary Authorities' provision for infectious diseases, 120, 115, 288; and mortuaries, 144.
Lockers, 88, 310, 312.
London hospitals, annual subscribers to, 38.
   — cost per bed at certain, 32.
   — funded interest of, 40.
London Homœopathic Hospital, cost of alcohol at, 95.
London Hospital, the, cost per bed at, 32; cost of alcohol at, 95.
Lying-in hospitals, large and small, 26.
   —   and cottage hospitals, 28.
   —   mortality of, 26.
Lynton District Cottage Hospital, history and cost of, 216.
Lytham Cottage Hospital, cost of alcohol at, 95; small wards at, 158.

McKinnell's ventilator, 167.
Maidenhead Cottage Hospital, 246.
Man attendant for cottage hospital, 185, 237.
Manchester, medical remuneration at, 47.
Manhole, 191.
   diagram of, 191.
Manure from earth-closets, 186, 188.
Mary Hewetson Cottage Hospital, 9.
Massachusetts, cottage hospitals in, 327.
Maternity cases, cottages for, 346.
Matrons in the United States, 341.
   — in small hospitals, 69.
Mattresses, hair, 303.
Medical attendants, remuneration of, 46, 122, 339.
   — case book, particulars to be given, 22.
   — case book, specimen page of, 358.
   — department in United States cottage hospitals, 339.
   — director, appointment of, 62.
   —   — duties of, 64, 67.
   — profession and cottage hospitals, 7, 47, 59, 111, 339.
   — profession and paying patients, 16, 125, 339.
   profession, intercourse amongst, 7, 60, 65.
   — staff, payment of, 46.
   —   and pauper patients, 59.
   consultation on outside cases, 60, 115.
   —   rules for, 67.
Medicine, administration of, to semi-conscious patient, 100.
Medicine bottles, brackets for, 311.

376

INDEX.

Medicines, regularity in giving, 83.
Metropolitan hospitals, cost per bed, 32;
cost per patient for alcohol, 95.
Middlesex Hospital, cost per bed at, 32.
— cost of alcohol at, 95.
Midwifery cases in cottage hospital, 26, 346.
Midwives, registration of, 27.
Milford (Staffs), Sister Dora Convalescent Hospital, 261.
Miller Hospital, cost of alcohol at, 95.
Milton Abbas Cottage Hospital, boarding system at, 111; midwifery cases at, 27.
Mining districts, size of cottage hospitals for, 149, 160.
Mirfield Memorial Hospital, 266.
Model forms letter of recommendation, 55; return of cases treated, 23, 24, 358; statement of accounts, 356.
Model pavilion plan, convalescent institution, 282.
Model plan, fever hospital, 284, 292.
— — general hospital, 282, 294.
— — Local Government Board fever hospitals, 286.
— — mortuary, 284.
— — pavilion hospital, 277.
— — permanent fever hospital, 285.
— — small cottage hospital, 280.
— alternative plans for small cottage hospitals, 290.
— villa hospital, 296.
" Modern " sanitary fittings, evils of, 4.
Mortality before the antiseptic era, 19.
Morton Hospital, Taunton, Mass., history of, 335.
Mortuaries, 142, 165.
— and infectious corpses, 145.
— cost of, 147, 284.
— Local Government Board, 145.
Moule's closets, 187.
Movable bath, 314.
— earth-closets for wards, 314.
Mustard poultices, 100.

Napper, the late Mr., the founder of cottage hospitals, 5; on advantages of cottage hospitals, 2, 5.
Necessaries for patients, 309.
New hospital, starting of, 15.
Newcastle Royal Infirmary, cost per bed at, 32; source of income, 35; results of major amputations at, 20.
Newington Hospital, closure of, 4.
Newton Hospital, 95, 114.
Northam Cottage Hospital for chronic cases, 240.
Northampton, patients' payments, 46.
— St. John's Hospital, 288.
North Ormesby Cottage Hospital, convalescents at, 132; claim to be first cottage hospital refuted, 5; history of, 10.
Nose, foreign bodies in, 102.

Nuisance from slop-water, 82.
Number of cottage hospitals, 6, 10, 17.
Nursing arrangements in large cottage hospitals, 158.
— arrangements in small cottage hospitals, 152.
— at Boston, 73.
— at Ditchingham, 70.
— at Middlesbrough, 70.
— at Walsall, 70.
— department, meagre information in Annual Report, 74.
— former systems, 78.
— form of certificate of training, 72.
— institutions attached to cottage hospitals, 70.
— of private cases in United States hospitals, 353.
— probationers in cottage hospitals, 71.
— qualifications required in cottage hospitals, 71, 78.
— systems in force in cottage hospitals, 68, 152.
Nurses, advice to, 78, 83.
and bad ventilation, 81.
— and medicine giving, 83.
— for cottage hospitals, 71.
— qualifications required in cottage hospitals, 71.
— training institutions for, 70.
— training institutions in the United States, 340, 343.

Oakum and tenax pads, 325.
Objections to cesspits, 183.
— to pavilion hospitals, 290.
— to ticket system, 53, 110.
Operating-room, 105, 164.
— table, 164.
Operations at cottage hospitals, 21.
Out-patients at Harrow Cottage Hospital, 212.
— department, 114.
— dinners for, 226.
— letters, number given, 114; time for which they last, 115.

Pads for ward purposes, 324.
Page, Dr. Frederick, statistics of amputations at Newcastle Royal Infirmary 20.
Painting and varnishing of walls, 117, 161, 233, 305.
Paper varnished for walls, 118, 308.
Parian cement for hospital walls, 117, 161.
Patients, admission of friends, 106.
— at Cranleigh in 1864, 5.
— at Leek, 210.
— average weeklycost in America, 337.
— boarded by nurse or matron, 111.
— cleanliness of, 85.
— clothes of, 88, 314.
— necessaries of, 309.
— payments, 6, 43, 212, 216.
— remunerative paying, 122.
— statements to be given in Annual Report, 23, 24.

Pauperising tendency of gratuitous medical relief, 45, 56.
Paupers, admission of, 51.
Pavilion hospitals, 160, 278.
Payment of medical staff, 40, 46.
Payments by patients, 6, 43, 56, 122, 212, 216, 217, 218, 219, 221, 225, 238, 241, 337.
Paying probationers, 73.
Percentage sources of income and of total income at cottage and general hospitals, 35.
Pedestal hygienic water-closet, 191.
Permanent fever hospital, model plan of, 285.
Petersfield Cottage Hospital, description of, 213, 249; cost of alcohol at, 95; case book, 358; mortuary at, 143.
Petworth Cottage Hospital, history and description of, 224.
Pictures in wards, 306.
Pins, safety, 325.
Pipes for sub-irrigation, 184, 198.
Plates, hot-water, 314.
Polished floor, 162.
Poplar Hospital, cost of alcohol at, 95.
Porter, employment of, 188, 237.
Portland cement for walls, 117, 161.
Poultices, 98.
Prayers in cottage hospitals, 84, 108.
Prescription paper, 311, 313.
Preservation of ice, 91.
Pridgin-Teale grate, 172.
Probationer nurses, 71.
Probationers in the United States, 343.
— paying, 73.
— payments at Boston Cottage Hospital, 73.
Provident dispensary system advocated for cottage hospitals, 46.
Provident dispensaries, 116.
— nature of cottage hospitals, 44, 47.
Provincial general hospitals, cost per bed at certain, 32.
Pumping water, 178.

Qualifications of nurses for cottage hospitals, 71, 78.

Rain-water supply, 179, 245.
Ratio of beds to population, 9, 10, 150.
Recommendation, letters of, 52, 55, 110, 114.
Redruth, convalescents at, 131.
Registration of midwives, 27.
Reigate Cottage Hospital, description and history of, 226.
Religious services, 84, 108, 334.
Removal of earth from closets, 185, 186.
Remuneration of medical staff, 40, 46.
-- at Northampton, 46.
-- at Manchester, 47.
Remunerative paying patients, 122.
— at Leek, 219.
Reserve capital, amounts of necessary, 41.
Retford Cottage Hospital, cost of alcohol at, 95.

Romford Cottage Hospital, cost of alcohol at, 95.
Roof space of cottage hospitals, 179.
Ross Memorial Hospital, 160.
Royal Free Hospital, cost of alcohol at, 95.
Rules, at Harrow Cottage Hospital, 211.
— draft for cottage hospitals, 365.
— for medical staff, 67, 212.
— Lynton District Cottage Hospital, 216.
Rutland, want of hospital accommodation in, 9.

Saffron Walden, patients' payments, 241.
St. George's Hospital, cost of alcohol at, 95.
St. Mary's Hospital, cost of alcohol at, 95.
St. Thomas's Hospital, pay wards at, 123.
Sand bags, 324.
Sanitary arrangements at Grantham Hospital, 244.
— at Stamford fever wards, 256.
Sanitary authorities and fever hospitals, 120.
— and mortuaries, 145.
Sanitary condition of wells, 178.
Sanitary fittings, the evils of "modern," 4.
Savernake Cottage Hospital, 114; mortuary at, 143.
Sawdust pads, 325.
Schede's amputation statistics, 20.
Scheme for foundation of cottage hospitals, 15.
Screens, 314.
Scullery, 164.
Seamen's Hospital, cost of alcohol at, 95.
Self-acting earth-closets, 187.
Separate wards necessary for midwifery patients, 27.
Sewage disposal, 184, 192.
— — at Eastwick, 200.
Shallow well water, 177.
Shedfield Cottage Hospital, kitchen at, 115.
Sheet, changing, 86.
Shepton Mallet Cottage Hospital, 95.
Sherborne, Yeatman Hospital, 238.
Sheringham valves, 167.
Shot bags, 321.
Silicate paint for hospital walls, 117.
Simon, Sir John, on pay wards at St. Thomas's Hospital, 124.
Single-bedded wards, 158.
Single-storied hospitals, 159.
Sink-water, disposal of, 190, 199, 202.
Sister Dora Convalescent Hospital, Milford (Staffs), 261.
Site of cottage hospitals, 153.
Size of cottage hospitals, 150.
— of wards, 160.
Slings for arms, 316.
Slops, disposal of, 82, 196, 199, 202, 204.
Slow combustion calorigen, 175.
Small cottage hospitals, model plan, 280.

25*

Smallpox hospitals, 288.
Soil pipes, 188, 190.
Southam Hospital, closure of, 11.
Spirits, consumption of, in cottage and general hospitals compared, 95.
Sponges not to be used for washing wounds, 105.
Sprained ankle, treatment of, 97.
Spread of the cottage hospital movement, 6, 9, 10.
Stamford fever wards, 252.
Steele, the late Dr. J. C., on hospital mortality, 19.
Storage tank for sewage, 192.
Stoves, calorigen, 172.
— Galton's, etc. (see Grates).
Straight model for cottage hospitals (Burdett's), 290.
Stretchers, 96.
Sub-irrigation, 184, 198.
Subscriber's letter of recommendation, 52, 55, 110, 114.
Subscriptions, annual, 36.
Sub-surface irrigation, 184, 198.
Success of treatment in large and small hospitals, 20.
Summary of patients to be given in Annual Report, 23.
Surgical cases before and during antiseptic era, 20.
Surgical excellence in cottage hospitals, 2.
Surgical hammock, 317.
        instruments, 299.
Surplus income, how to deal with it, 48.
Sutton, mortuary at, 143.
Swete, Dr. Horace, on cottage hospitals, 3; on ambulances, 321.
Synnot, Miss, and boarding out children in farm-houses, 138.
Syphon tank, 193.
Systems of nursing in force, 68.

Table for ward, 314.
Tamworth Cottage Hospital, cost of alcohol at, 95.
Tanks for sewage, 192, 193.
Taylor, Mr., plan of excrement disposal, 193.
Teale grates, 172.
Temperature of wards, 81.
Tenby, convalescents at, 131.
Testing drains, 180.
Tetbury Cottage Hospital, cost of alcohol at, 95.
Thermometer, mode of using a clinical, 98.
Thermoson grate, 170.
Ticket system, objections to, 53, 110.
Towels, 320.
Towel roller on locker, 311.
Tow pads, 325.
Trained nurses, 70.
Training institutions for nurses, 70.
        — in the United States, 342, 343.
Traps to sewers, 189, 190.
Treatment of accidents on admission, 95.

Treatment of burns, 103.
    · of fainting, 101.
    — of fits, 102.
    — of foreign bodies in ear, etc., 102.
    — of frost-bite, 102.
    — of hæmorrhage, 103.
    — of injuries of head, 104.
    — of sprained ankle, 97.
Trusses for hospital patients, 241.
Tube wells for water supply, 177.

Ulverston, mortuary at, 143.
Uneaten food, 88.
Uniformity of accounts, 30, 356.
United States, Cottage Hospitals in. (See America).
University College Hospital, cost of alcohol at, 95.
Urban districts and infectious hospitals, 120.
Urinals, 156, 195.

Varnished papers for hospital walls, 118, 308.
Varnishing of walls, 117.
Ventilating grates, 170.
Ventilation, 80, 166.
    - - of Boston Hospital, 233.
    — of cisterns, 179.
        of closets, 156.
    - - of drains, 189, 245.
    — of kitchen, 157.
    — of manholes, 190.
        of Sheringham valves, 167.
    — of underfloors, 162.
    · of wards, 80.
Ventilators, McKinnell's, 167.
Vertical ventilation, 168.
Visits of friends to patients, 106.

Walker Hospital, description and history of, 219; mortuary at, 143.
Walls, damp, 157.
    · · of wards, 307.
    — painting and varnishing, 117, 161, 308.
    — papering and varnishing, 118, 308.
Walsall Cottage Hospital, 184.
Ward furniture, 312.
    — single-bedded, 158.
Wards, air space of, 166.
    — bad arrangement of, 154.
    — for paying patients, 122.
    — size of, 166.
    — ventilation of, 81, 166.
Warming of wards, 166.
Waring, Dr., on cottage hospitals, 3.
Wash-hand basin for surgeon, 320.
Washing patients, 85.
Watford Cottage Hospital, 264.
Water-closets, 82, 183, 188, 191.
Water supply, 176.
    —   — of bath-rooms, 179.
    —   — of mining districts, 180.
    ·   — of urinals, 195.
    ·   tube wells, 177.
        hot, system, 182.

Water, softening of, 181.
— waste preventers, 188.
Wells, Abyssinian tube, 177.
— ordinary, points to be remembered in construction of, 178.
Wellington Cottage Hospital, cost of alcohol at, 95.
West London Hospital, cost of alcohol at, 95.
Westminster Hospital, cost per bed at, 32; sources of income, 35; cost of alcohol at, 95.
West Cornwall Miners' Hospital, convalescents at, 131; history and description of, 221.
Weston Favell Convalescent Institution, 282.

Wilson, Dr. George, on water supply, 177.
Window ventilation, 81, 166, 233.
Wines and spirits, consumption of, in cottage hospitals, 95.
Wine and beer cellar, 163.
Winter, ventilation during, 169.
Wirksworth Cottage Hospital, 41, 225.
Wood Green Cottage Hospital, 274.
Wringing machines for fomentation cloths, 100.
Wrington Hospital, closure of, 13.
Wynter, Dr., on cottage hospitals, 2.

Yate Hospital, closure of, 13.
Yeatman Hospital, Sherborne, history and description of, 238.

ABERDEEN UNIVERSITY PRESS

CATALOGUE

OF WORKS ON

MEDICINE, SCIENCE

HOSPITALS AND HOSPITAL

CONSTRUCTION

HYGIENE, NURSING, ETC.

PUBLISHED BY

THE SCIENTIFIC PRESS, LIMITED

28 AND 29 SOUTHAMPTON STREET, STRAND, LONDON, W.C

INDEX TO SUBJECTS.

	PAGE		PAGE
CHILDREN, DISEASES AND MANAGEMENT OF, ...	3	W.T. KEENER CO.'S PUBLICATIONS,	10
DIET, ...	3	MEDICINE, SURGERY, CHEMISTRY, etc., ...	11
GENERAL, ...	4	NURSING, TEXT-BOOKS ON ...	13
HOSPITALS AND HOSPITAL CONSTRUCTION, ...	4	OBSTETRICS, ...	12
HOSPITAL ACCOUNTS, ...	6	PERIODICALS, ...	16
HYGIENE, ...	8	SCIENCE PROGRESS, ...	16
INSANITY, ...	8	THE HOSPITAL, ...	16
JOHNS HOPKINS HOSPITAL PUBLICATIONS,...	9	BURDETT'S HOSPITALS AND CHARITIES, ...	16
JOHNS HOPKINS HOSPITAL REPORTS, ...	10	BURDETT'S OFFICIAL NURSING DIRECTORY, ...	16

LATEST PUBLICATIONS.

" THE BURDETT SERIES " OF POPULAR TEXT-BOOKS ON NURSING.
Small crown 8vo, cloth. Price 1s. each.

No. 1.—**Practical Hints on District Nursing.** By AMY HUGHES, Superintendent of The Nurses' Co-operation; late Superintendent of the Central Training Home, Queen Victoria's Jubilee Institute for Nurses for the Sick Poor, and of the Metropolitan and National Nursing Association.

No. 2.—**The Matron's Course.** An Introduction to Hospital and Private Nursing. By Miss S. E. ORME, Lady Superintendent, London Temperance Hospital.

No. 3.—**The Midwives' Pocket Book.** By HONNOR MORTEN, Author of *How to Become a Nurse, and How to Succeed; The Nurse's Dictionary*, etc., etc.

FORTHCOMING BOOKS.

A Manual of Hygiene for Nurses and Students. By JOHN GLAISTER, M.D., D.P.H. (Camb.). Crown 8vo, profusely illustrated with over 70 drawin s, cloth, 3s. 6d.

The Mystery and Romance of Alchemy and Pharmacy. By C. J. S. THOMPSON, Author of *The Chemists' Compendium, The Cult of Beauty*, etc., etc. Crown 8vo, cloth gilt, profusely illustrated, 5s.

Dr. Mendini's Hygienic Guide to Rome. Translated from the Italian, and edited with an additional chapter on Rome as a Health Resort. By JOHN J. EYRE, M.R.C.P., L.R.C.S., Ireland. Crown 8vo, cloth, 2s. 6d. net.

Fever Nursing. By William Harding, M.D. Ed., M.R.C.P. Lond. Author of *Mental Nursing*, etc. Small Crown 8vo, cloth, 1s.

Child Life under Queen Victoria. By Mrs. FURLEY SMITH. 1s.

July, 1897.

CATALOGUE

OF WORKS PUBLISHED BY

THE SCIENTIFIC PRESS, Ltd.,

28 AND 29 SOUTHAMPTON STREET, STRAND, LONDON, W.C.

CHILDREN, DISEASES AND MANAGEMENT OF.

Spinal Curvature and Awkward Deportment: their Causes and Prevention in Children. By Dr. GEORGE MÜLLER, Professor of Medicine and Orthopœdics, Berlin. English Edition, edited and adapted by RICHARD GREENE, F.R.C.P., Ed. Crown 8vo, cloth gilt, with many Plates and Illustrations, 2s. 6d.

"Dr. Müller's little book is an able essay upon the effect produced by exercise, attitude, and movement upon the growth and development of the body. He further gives directions which any intelligent parent or teacher could carry out with the help of very simple apparatus, showing how to avoid, or, in early cases, to cure certain malformations, which, in the majority of cases, arise either from the neglect of very simple hygienic rules or from carelessness."—*Daily Chronicle.*

The Mother's Help and Guide to the Domestic Management of Her Children. By P. MURRAY BRAIDWOOD, M.D., Formerly Senior Medical Officer to the Wirral Hospital for Sick Children. Second Edition, with Addenda, crown 8vo, cloth, 2s.

"The book is written plainly and without pedantry, and the directions given for the management of the child are sensible and discreet. . . . . We do not doubt the need of it in the ignorance of many mothers, and we can recommend it as likely to enlighten that ignorance."—*Glasgow Medical Journal.*

"This little book is admirably conceived and executed . . . . to many young mothers it should prove invaluable."—*Provincial Medical Journal.*

"We can confidently recommend the book for the purpose for which it is intended, and it contains many hints that would be valuable to the junior practitioner."—*Bristol Medico-Chirurgical Journal.*

DIET.

Infant Feeding by Artificial Means: A Scientific and Practical Treatise on the Dietetics of Infancy. By S. H. SADLER, Author of "Suggestions to Mothers," "Management of Children," "Education". Second Edition, with a new chapter on the history of infant feeding by artificial means in the early ages, illustrated with coloured and other plates. Facsimile Autograph Letters from the late Sir ANDREW CLARK, M. PASTEUR, etc., etc. Crown 8vo, cloth gilt, with 17 Plates and many Illustrations in the text, 5s.

"Mrs. Sadler's book deals with the question of the artificial feeding of infants, and contains a very useful collection of the views of the best-known English authorities upon the subject. The truly terrible ignorance displayed by mothers, especially amongst the poor, upon the subject of infant feeding is answerable for an infant mortality so great as to be appalling."—*Daily Chronicle.*

DIET—(*Continued*).

Diet in Sickness and in Health. By Mrs. ERNEST HART,

Bachelier-ès-Sciences-ès-Lettres (restreint), formerly Student of the Faculty of Medicine of Paris, and of the London School of Medicine for Women. With an Introduction by Sir HENRY THOMPSON, F.R.C.S., M.B., Lond. Fourth Thousand. Demy 8vo, blue buckram, gilt, with numerous illustrations, 3s. 6d.

Sir Henry Thompson, in his Introduction, says:—" I do not hesitate to express my opinion that the present volume forms a handbook to the subject thus briefly set forth in these few lines, which will not only interest the dietetic student but offer him, within its modest compass, a more complete epitome thereof than any work which has yet come under my notice ".

" Mrs. Hart speaks not only with the authority derived from experience, but with the ease and freedom of an expert. We have perused this book with great pleasure, and feel sure that many will find in it much to lighten the days of those whose digestion is enfeebled."—*The Hospital.*

Art of Feeding the Invalid. By a MEDICAL PRACTITIONER

and a LADY PROFESSOR OF COOKERY. Demy 8vo, cloth gilt, 260 pp., 3s. 6d.

" This is a useful book. . . . To the housekeeper who has a dyspeptic, gouty, or diabetic member in her family, this book cannot fail to be of great value, and save her much anxious thought, and prevent her making serious mistakes."—*British Medical Journal.*

GENERAL.

Heavy Trial Balances made Easy. A new method to

secure the immediate agreement of Trial Balances without trouble. By J. G. CRAGGS, F.C.A. Royal 8vo, illustrated by coloured examples. Cloth, price 2s. 6d.

Charity Organisation and Jesus Christ. By Rev. C.

L. MARSON, M.A., Perpetual Curate of Hambridge. Price 1s.

This tractate is a plea for the needy and against the niggardly. It is addressed to such men and women as think that Scientific Almsgiving, which conflicts with Scientific Theology or with Redeemed Humanity, had better get out of the way as soon as possible.

HOSPITALS AND HOSPITAL CONSTRUCTION.

Cottage Hospitals, General, Fever, and Convales-

cent. Their Progress, Management, and Work in Great Britain, Ireland, and the United States of America With an alphabetical list of every Cottage Hospital at present opened. By Sir HENRY C. BURDETT, K.C.B., Author of " Hospitals and Asylums of the World," etc., etc. Third Edition, profusely Illustrated, with nearly 50 Plans, Diagrams, etc., including a Portrait of Albert Napper, Esq., the founder of Cottage Hospitals, crown 8vo, cloth gilt, 10s. 6d.

The present issue, although a Third Edition, is really a new book on the subject. During the fifteen years which have elapsed since the Second Edition was brought out, Cottage Hospitals have more than doubled in importance and number, so that the author has had practically to re-write the work. The changes which have taken place, and the opinions of experts, are well brought out, especially in the chapters dealing with construction, where a large number of new plans will be found.

"Contains all the information which could be required by any one who undertakes to found or to manage a Cottage Hospital."—*British Medical Journal.*

THE SCIENTIFIC PRESS, LTD., 28 AND 29 SOUTHAMPTON STREET, STRAND, W.C.

HOSPITALS AND HOSPITAL CONSTRUCTION—(Continued).

Hospitals and Asylums of the World.

Their Origin, History, Construction, Administration, Management, and Legislation; with Plans of the chief Medical Institutions, accurately drawn to a uniform scale, in addition to those of all the Hospitals of London in the Jubilee year of Queen Victoria's reign. By Sir HENRY C. BURDETT, K.C.B., formerly Secretary and General Superintendent of the Queen's Hospital, Birmingham; a Registrar of the Medical School; the " Dreadnought " Seamen's Hospital, Greenwich; Founder of the Home Hospitals Association for Paying Patients, the Hospitals Association, and the Royal National Pension Fund for Nurses; Author of " Pay Hospitals of the World," " Hospitals and the State," "Cottage Hospitals: General, Fever, and Convalescent," " The Relative Mortality of Large and Small Hospitals," Burdett's " Hospitals and Charities," etc. In Four Volumes, with a Portfolio of Plans, cloth extra, bevelled. Royal 8vo. Top Gilt. Price as under.

IN FOUR VOLUMES, AND A SEPARATE PORTFOLIO CONTAINING SOME HUNDREDS OF PLANS.

Price of the Book, complete, - - - - -	£8	8	0
Vols. I. and II. (only)—Asylums and Asylum Construction, - - - - - - -	4	10	0
Vols. III. and IV.—Hospitals and Hospital Construction,—with Portfolio of Plans, - - -	6	0	0
The Portfolio of Plans (20 ins. by 14 ins.), separately, price - - - - - - - -	3	3	0

SPECIAL NOTE TO BOOK BUYERS, LIBRARIANS, AND THE TRADE.

Only Five Hundred copies of this work were printed for sale, and Four Hundred copies are already disposed of.

"Mr. Burdett's monumental work on Hospitals."—*The Times.*

" Mr. Burdett is to be congratulated upon the completion of this monumental work: a lesser man would have been buried under the mass of his material. He has produced a book which should be found in all asylums, hospitals, and public libraries, and which every architect and medical man who aspires to an understanding of the principles of hospital construction, organisation, and management will do well to procure for himself. All honour to Mr. Burdett that he has erected a monument more enduring than brass." —*The British Medical Journal.*

'These magnificent volumes . . . the outcome of a vast amount of laborious investigation and of many journeys in Europe, America, and the British Colonies. . . ."—*The National Observer.*

"The most exhaustive work on the subject extant. It is full of research, historical and medical, referring to hospitals and asylums, and must remain for a considerable time the leading book of reference to the classes of institution it deals with. Forming in itself a very valuable work of reference to the architect and all managers of hospitals."—*The Building News.*

It would be impossible to exaggerate the value and interest of the work." *The Saturday Review.*

"Comprehensiveness of scope and thoroughness of treatment."—*The Glasgow Herald.*

THE SCIENTIFIC PRESS, LTD., 28 AND 29 SOUTHAMPTON STREET, STRAND, W.C.

HOSPITALS AND HOSPITAL CONSTRUCTION—(Continued).

Hospitals, Dispensaries, and Nursing. Being Transactions of Section III. International Congress of Charities, Correction, and Philanthropy, held in Chicago, 1893. Royal 8vo, 500 pp., 60 Illustrations, cloth gilt, 21s. net (see also p. 9).

Burdett's Hospitals and Charities. Being the Year Book of Philanthropy. Containing a Review of the Position and Requirements, and Chapters on the Management, Revenue and Cost of the Charities. An Exhaustive Record of Hospital Work for the Year. It will also be found to be the most useful and Reliable Guide to British, American, and Colonial Hospitals and Asylums, Medical Schools and Colleges, Religious and Benevolent Institutions, Dispensaries, Nursing and Convalescent Institutions. Edited by Sir HENRY C. BURDETT, K.C.B., Author of "Hospitals and Asylums of the World," etc., etc. Published annually, about 1000 pp., crown 8vo, cloth gilt, 5s.

"Still remains the standard work of reference upon all points which relate to charities, British, American, and Colonial."—*Daily Telegraph.*

"Every year this admirable compendium has grown more valuable as an authoritative and comprehensive work of reference."—*The Speaker.*

Suffering London: or, the Hygienic, Moral, Social and Political Relations of our Voluntary Hospitals to Society. By A. EGMONT HAKE, with an Introduction by Sir WALTER BESANT. Demy 8vo, 250 pp., cloth boards, 3s. 6d.

A graphic representation of the sufferings alleviated by our Hospitals.

"Is a powerful and eloquent plea for the more generous support, by those who have the means, of institutions which have grievously suffered of late by the diversion of public charity towards objects which are less indisputably worthy in themselves, and less demonstrably necessary to the welfare of our social organism. . . . Has our cordial sympathy."—*The Times.*

The Furnishing and Appliances of a Cottage Hospital. By the Honourable SYDNEY HOLLAND. Price 6d.

HOSPITAL ACCOUNTS.

The Uniform System of Accounts, Audit, and Tenders, for Hospitals and Institutions, with Certain Checks upon Expenditure; also the Index of Classification. Compiled by a Committee of Hospital Secretaries, and adopted by a General Meeting of the same, 18th January, 1892. And certain Tender and other Forms for securing Economy. By Sir HENRY C. BURDETT, K.C.B., Author of "Hospitals and Asylums of the World," "Burdett's Hospitals and Charities," etc., etc. Demy 8vo, profusely illustrated with Specimen Tables, Index of Classification, Forms of Tender, etc., cloth extra, 6s.

"Hospital secretaries and book-keepers will find a very valuable help in the wearisome business of arranging their figures according to this system in Mr. Henry Burdett's clear and concise treatise. He not only guides his readers through the maze of the official form of balance-sheet, but also provides specimen pages of books suitable for large hospitals, which will facilitate the final classification of accounts at the end of the year. Will be welcomed by every hospital secretary and auditor."—*Morning Post.*

THE SCIENTIFIC PRESS, LTD., 28 AND 29 SOUTHAMPTON STREET, STRAND, W.C.

HOSPITAL ACCOUNTS—*(Continued).*

Account Books for Institutions designed in accordance with the Uniform System of Accounts for Hospitals and Institutions. By Sir HENRY C. BURDETT, K.C.B., Author of "Hospitals and Asylums of the World," "Burdett's Hospitals and Charities".

Being a complete set of Account Books, ruled in accordance with the Uniform System adopted by the Metropolitan Hospital Sunday Fund. Designed and constructed for the convenience and assistance of Secretaries who present annual returns for participation in the Hospital Sunday Fund Grants.

These Books are ruled so that the various descriptions of Receipts and Expenditure, etc., etc., may be entered uniformly under the special headings and in the columns prepared for them.

By their use the labour of Hospital Secretaries is reduced to a minimum ; and the Accounts of all Public Institutions can be kept in a uniform manner.

To facilitate the compilation of statements for the purposes of the Metropolitan Hospital Sunday Fund, it is necessary to use a complete set of Account Books, ruled so that the ordinary daily entries may give the totals required for such statements.

These Account Books have been prepared to supply this need. Every detail has been minutely dealt with, and all possible entries provided for. Printed headings and divisions render the system perfectly plain, and lessen the work of keeping the accounts in a very large degree.

(For description of books and prices, see below, also next page.)

Series No. I. For Hospitals and Institutions. (Any of these Books may be purchased separately ; a reduction is made if the complete set is taken.)

1. **Analysis Journal.** Royal, 350 pp. Divided into seven sections, with parchment tags. Headings as follows :
   A. Maintenance.
   1. Provisions.
   2. Surgery and Dispensary.
   3. Domestic.
   4. Establishment Charges.
   5. Rent.
   6. Salaries.
   7. Miscellaneous Expenses.
   B. Administration.
   1. Management.
   2. Finance.
   (Uniform with Headings in Income and Expenditure Table.)
   Each Heading is ruled with pages containing 15 columns. Price **£3.**
2. **Cash Book.** Foolscap, 300 pp., printed Headings, specially ruled. Price **£1.**
3. **Cash Analysis and Receipt Book.** Foolscap, extra width, 300 pp., ruled, with 15 columns, printed Headings identical with Income and Expenditure Table. Price **£1. 5s.**
4. **Secretary's Petty Cash Book.** 300 pp., foolscap, ruled with 13 columns, printed Headings identical with Income and Expenditure Table. Price **£1.**

5. **List or Register of Annual Subscribers.** Double foolscap, 300 pp., ruled with 10 columns, printed Headings for the years 1893-1900 inclusive. Price **£1. 15s.**
6. **Alphabetical Register Book.** F'cap., 300 pp., ruled, with date, name and address, and cash columns, cut into alphabetical sections, with totals at end for annual balances. Price **£1. 6s.**
7. **Subscription Register.** Foolscap, ruled with 8 columns, printed Headings for date when subscriptions become due, name and address, and cash columns for years 1893 to 1898 inclusive. Price **£1. 1s.**
8. **Linen Register.** Foolscap long quarto, 300 pp., ruled with columns for Stock, Condemned, Remaining and Issued Linen, also total in Ward and Remarks, also printed lists of Hosp. Linen. Price **8s. 6d.**
9. **Income and Expenditure Table.** Basis of uniform system. For yearly balances and statements. Royal, Price 6d. per sheet, or 5s. per dozen.
10. **Special Appeal Account Table.** Foolscap. Price 3d. per sheet, or 2s. 6d. per dozen.

Price of Set complete, £10 net.

(For Series No. 2, for Cottage Hospitals, see next page.)

THE SCIENTIFIC PRESS, LTD., 28 AND 29 SOUTHAMPTON STREET, STRAND, W.C.

HOSPITAL ACCOUNTS—(Continued).

Series No. 2. For Cottage Hospitals and Smaller Institutions.

1. Cash Analysis, Receipt and Expenditure Book. Double foolscap, 300 pp., ruled with 18 columns, printed Headings identical with Income and Expenditure Tables, specially prepared for Cottage Hospitals and small Institutions. **Price £1. 15s.**

2. Secretary's Petty Cash Book. 300 pp., foolscap, ruled with 13 columns, printed Headings identical with Income and Expenditure Table. **Price £1.**

3. Income and Expenditure Table. Basis of uniform system. For yearly balances and statements. Royal. Price **6d.** per sheet, or **5s.** per dozen.

4. Linen Register. Foolscap, 300 pp., ruled with columns for Stock, Condemned, Remaining, and Issued Linen, also total in Ward and Remarks, also printed Lists of Hospital Linen. **Price 8s. 6d.**

5. List or Register of Annual Subscribers. Foolscap. 300 pp., ruled with columns for the years 1893-1900 inclusive. **Price £1.**

6. Special Appeal Account Tables. Foolscap. Price **3d.** per sheet, or **2s. 6d.** per dozen.

Price of Set complete, £4 net.

The books are all uniformly bound in half basil, with gilt letterings, and are ruled on best paper, and in every way prepared for practical use at the offices of Institutions.

The Cottage Hospital Case Book, or, Register of Patients.

Prepared and ruled in accordance with what has been found in practice to be the most approved system, and suitable for large or small Cottage Hospitals. Printed on best superfine account-book paper, and bound in half basil, green cloth.

In Two Sizes:—

No. I. containing space for about 150 in-patient cases, with rulings for 8 weeks' payments to each case. £0 10 0

No. II. containing space for about 300 in-patient cases, with rulings for 8 weeks' payments to each case. 0 12 6

HYGIENE.

Helps in Sickness and to Health: Where to go and What to Do.

Being a Guide to Home Nursing, and a Handbook to Health in the Habitation, the Nursery, the Schoolroom, and the Person, with a chapter on Pleasure and Health Resorts. By Sir Henry C. Burdett. K.C.B., Author of " Hospitals and Asylums of the World," etc. Crown 8vo, 400 pp., Illustrated, cloth gilt, 5s.

" It would be difficult to find one which should be more welcome in a household than this unpretending but most useful book."—*The Times.*

" No medical or general library can be complete without such a book of ready reference."—*Lancet.*

INSANITY.

Outlines of Insanity.

A Popular Treatise on the Salient Features of Insanity. By Francis H. Walmsley, M.D., Medical Superintendent of the Darenth Asylum, Member of Council of Medico-Psychological Association. A clearly written manual in which the various features of mental disorder are fully expounded. The author's style is interesting and the reader's attention is riveted throughout. The Popular Edition is well suited to nurses. Demy 8vo, cloth extra, 3s. 6d. Popular Edition. stitched. 1s. 6d.

" The work is accurate, and it is well arranged and pleasantly written. It will be serviceable to those for whose use it has been designed."—*British Medical Journal.*

THE SCIENTIFIC PRESS. Ltd., 28 and 29 Southampton Street, Strand, W.C.

INSANITY—(Continued).

Mental Nursing. A Text-book specially designed for the instruc tion of Attendants on the Insane. By WILLIAM HARDING, M.D. Ed., M.R.C.P. Lond. Second Edition, enlarged, demy 8vo, profusely illustrated, nearly 200 pp., cloth, 2s. 6d.

The want of a complete book for the instruction of Asylum Attendants has been long felt, and it is with confidence that the publishers recommend this work to the managers of all Institutions for the Insane.

"Nothing could be better devised to serve as a text-book. The lectures are so simple and so instructive, that they cannot fail to impress and interest readers. . . . Of great value to all nurses. . . . No institution should fail to supply their officials with copies of the work."—*Local Government Journal.*

THE JOHNS HOPKINS HOSPITAL PUBLICATIONS.

Hospitals, Dispensaries, and Nursing. Being Trans- actions of Section III. International Congress of Charities, Correc- tion and Philanthropy, held in Chicago, 1893. Royal 8vo, 500 pp., 60 Illustrations, cloth gilt, 21s. net.

"This work is really an encyclopædia in one volume, and contains a vast amount of information of all kinds. No physician who desires to have at hand a ready reference book concerning hospitals, etc., should fail to secure this volume."—*New York Medical Record.*

The Organisation of Charities. Being Abstracts of the Discussion on the Subject at Chicago, 1893. A valuable collection of contributions on this important question by leading writers in the United States, Great Britain, and the Continent of Europe. The authors' names comprise Rev. F. G. Peabody, Mrs. Chas. R. Lowell, Miss Frances Morse, Rev. B. H. Alford, Rev. E. H. Bradly, D.D.; Rev. Brooke Lambert, Baron von Reitzenstein, Messrs. C. N. Nichol- son, C. P. Larner, T. Mackay, A. McDougall, R. A. Leach, George Milne, H. Willink, Baldwin Fleming, C. H. Wyatt, Miss Sharpe Miss Sturge, Miss Prideaux, etc., etc. Cloth gilt, 8vo, 375 pp., 6s. net.

"All practical philanthropists and especially all who are engaged in that newest and most practical form of it, which is now called 'Charity Organisa- tion,' ought to procure the volume and study it carefully. . . . We must . . . commend this most interesting work to their serious perusal."—*British Medical Journal.*

Report of the Proceedings of the International Congress of Charities, Correction, and Philan- thropy. Held in Chicago, 1893. I. General Exercises, etc. II. The Public Treatment of Pauperism. Edited by JOHN H. FINLEY, Ph.D., President of Knox College. Royal 8vo, cloth, 320 pp., 6s. net.

Commitment, Detention, Care and Treatment of the Insane. Being a Report of the Fourth Section of the International Congress of Charities, Correction, and Philanthropy, held in Chicago, 1893. Edited by G. ALDER BLUMER, M.D., Super- intendent of Utica State Hospital, and A. B. RICHARDSON, M.D., Superintendent of Columbus Asylum for Insane. This Volume also contains a Report on the Care and Training of the Feeble-minded, by GEORGE H. KNIGHT, M.D.: and on the Prevention and Repression of Crime, by FREDERICK H. WINES, LL.D. Royal 8vo, cloth, 320 pp., 6s. net.

THE JOHNS HOPKINS HOSPITAL REPORTS.

The Scientific Press have been appointed sole agents for the sale of the above Reports in the United Kingdom. The undermentioned volumes (with the exception of Vol. I.) are stocked at 28 and 29 Southampton Street, Strand, London, W.C., and can be supplied at once.

The Johns Hopkins Hospital Reports. VOLUME I.
Now ready. 4to, 423 pp. and 99 Plates. Price 21s. net.

This Volume will not be sold separately, but only to Subscribers to the Series as far as issued (Vols. I.-VI.).

VOLUME II. 4to., 570 pp., 28 Plates and many Illustrations. Price 21s. net, bound in cloth. Its contents are as follows : --
Medical Report for 1890, I.—Medical Report for 1890, II.—Report in Gynæcology, I.—Report in Surgery, I.—Report in Neurology, I.—Report in Pathology, I.

VOLUME III. 4to. 766 pp., 69 Plates and many Illustrations. Price 21s. net., bound in cloth. Its contents are as follows :—
Report in Pathology, II.—Report in Pathology, III.—Report in Gynæcology, II.

The Report in Gynæcology II., of Vol. III., can also be obtained separately, price 12s. 6d. net.

VOLUME IV. 4to, 504 pp., 22 Plates and many Illustrations. Price 21s. net, bound in cloth. Its contents are as follows :—
Reports on Typhoid Fever.—Neurology, II.—Surgery, II.—Gynæcology, III.—Pathology, IV.

All the Reports of Volume IV. can be obtained separately.

VOLUME V. 4to, 490 pp., 32 Plates and Charts, price 21s. net. Its contents are as follows :—
The Malarial Fevers of Baltimore.—A Study of Some Fatal Cases of Malaria. Studies in Typhoid Fever.

The Reports on The Malarial Fevers of Baltimore, and a Study of Some Fatal Cases of Malaria, of Vol. V., can be obtained separately.

VOLUME VI. (1897) is now in progress. Subscription for the vol., 21s. net.

List with complete contents of volumes published to date sent on application.

THE W. T. KEENER CO.'S PUBLICATIONS.

The Scientific Press beg to announce that they have been appointed agents in Great Britain for the sale of the publications of the W. T. Keener Co., of Chicago, and have now the following new works in stock :—

Experimental Surgery. N. SENN. M.D. 8vo, cloth, Illustrations, 20s. net.

Stricture of the Urethra. G. F. LYDSTON, M.D. 8vo, cloth, Plates and Illustrations, 12s. net.

Varicocele and its Treatment. G. F. LYDSTON, M.D. 8vo, cloth, Illustrations, 5s. net.

Electricity in Diseases of Women and Obstetrics. By FRANKLIN H. MARTIN, M.D. Illustrations, Royal 8vo, cloth, gilt, 8s. net.

Intestinal Surgery. N. SENN, M.D. 8vo, cloth, Illustrations, 10s. net.

Etiology of Osseous Deformities of the Head, Face, Jaws, and Teeth. E. S. TALBOT, M.D. Third Edition. 8vo, cloth, 461 Illustrations, 16s. net.

Rectal and Anal Surgery. E. ANDREWS, M.D. Third Edition. 8vo, 6s. net.

Technique of Post-mortem Examinations. L. HEKTOEN, M.D. 41 Illustrations, 7s. net.

Chemical Analysis for beginners. RÜDORFF, GIDSON, and MEAZEL. 8vo, cloth, 4s. net.

MEDICINE, SURGERY, CHEMISTRY, etc.

Medical History from the Earliest Times. By E. T.
Withington, M.A., M.B., Oxon. Demy 8vo, with two Plates, cloth gilt, over 400 pp., 12s. 6d. net.

"Medical history, as a branch of professional knowledge, has been too long a study by itself. . . . Medicine alone among the sister professions permits its votaries to leave its halls without that study of its past which 'intelligent curiosity' alone would of itself seem to impose as at once a pleasure and a duty. Of these works (i.e., treatises on medical history), by far the best is Dr. Withington's. He has come to his task well equipped with the necessary qualifications, professional, scholarly, and literary. His 'standard' is high, quite in keeping with that of the University whose function was declared by one of its most eminent sons to be this :—'To teach the student to know what knowledge of a subject means ;' and he keeps throughout a watchful eye on what it is necessary his reader should be familiarised with, and what he may, without prejudice, leave on one side. It is a history of medicine, in short, which the profounder student may accept as an introduction to the more exhaustive works of Continental, especially German, authors, and which the busy practitioner will find supplies him with all he may reasonably be expected to know. To have hit this happy mean has cost Dr. Withington no small pains, evidenced in the scrupulous care with which he dwells on the central figures of his subject, and relegates their satellites to compendious summaries, which he calls 'Notes' for those who are content with nothing but the *omne scibile* of the subject. These notes consist largely of references to original and more voluminous authorities, and if followed up will place the reader in full enjoyment of the 'atmosphere,' so to speak—the environment, moral and material—in which the medical world of the special period lived and moved. . . . But we have said enough to attract the reader to a work as interesting as it is attractive, and will merely add that the illustrations form a highly effective feature of some of its obscurer passages, and that an elaborate, though not too minutely detailed, index, facilitates references admirably."—*Lancet.*

Lectures on Genito-Urinary Diseases. By J. C.
Ogilvie Will, M.D., C.M., F.R.S.E., Consulting Surgeon to the Aberdeen Royal Infirmary, and Examiner in Surgery in the University of Aberdeen. Demy 8vo, profusely illustrated with Coloured and other Plates and Drawings, 6s. net.

Contents:—Urethral Fever and Catheter Fever—Treatment of Retention of Urine—Gleet and its Treatment—On Varicocele—On Hydrocele—The Treatment of Syphilis—Appendix—Prescriptions for Syphilis, etc., etc., etc.

"We have no hesitation in recommending Dr. Ogilvie Will's work to the practitioner and student of medicine."—*Practitioner.*

Clinical Diagnosis: A Practical Handbook of Chemical and
Microscopical Methods. By W. G. Aitchison Robertson, M.D., F.R.C.P. (Edin.), Author of "On the Growth of Dentine," "The Digestion of Sugars," etc. Foolscap 8vo, cloth gilt, profusely illustrated, 6s.

"Students and practitioners ought to welcome heartily a handbook like the one before us, which contains the facts and nothing else, which without any pretence of literary veneer accurately and succinctly offers to one who is about to undertake the chemical or microscopical investigation of any specimen or any organ, just those points which he ought to know, describes exactly the methods he ought to pursue, mentions the things he ought to have at hand, and, if we may be allowed the expression, gives a recipe for the making of the investigation."—*The Hospital.*

THE SCIENTIFIC PRESS, Ltd., 28 and 29 Southampton Street, Strand, W.C.

MEDICINE, SURGERY, CHEMISTRY, etc. —*(Continued)*.

Myxœdema: and the Effect of Climate on the Disease. By A. MARIUS WILSON, M.D., B.S., L.R.C.P., Lond.; M.R.C.S., Eng. Foolscap 8vo, cloth gilt, 2s.

A valuable little treatise on a subject about which comparatively little has been written. The special feature of this work is its treatment of the effects of climate upon the disease.

" In this brief monograph the author outlines the features of Myxœdema as at present understood."—*New York Medical Journal.*

A New Method of Inhalation for the Treatment of Diseases of the Lungs. By W. H. SPENCER, M.A., M.D., Cantab.; M.R.C.P. London. Consulting Physician to the Bristol Royal Infirmary, etc., etc. Crown 8vo, cloth, 1s. 6d.

"In this little book Dr. Spencer discusses the usual methods of introducing drugs into the respiratory passages, and describes a method of using certain volatile oils and other antiseptic substances, so as to impregnate the air of a room with the drug in a state of vapour. The publication of the present brochure will, no doubt, induce many practitioners to give a more extended trial to what may prove a valuable means of treatment."—*Manchester Medical Chronicle.*

Surgical Ward Work. By ALEXANDER MILES, M.D., Edin.; C.M., F.R.C.S.E. A practical manual of clinical instruction for Students in the Wards. Concisely, simply, and comprehensively treated. Contents: Section I.—Antiseptic Surgery. Section II.—The use of Rest in Surgery. Section III.—Bandaging. Section IV.—Surgical Instruments and Appliances. Demy 8vo, cloth boards, copiously illustrated, 3s. 6d.

" The book fills a distinct hiatus in surgical literature."—*Glasgow Medical Journal.*

The Schott Treatment for Chronic Heart Diseases. By RICHARD GREENE, F.R.C.P., Ed. Price 6d.

OBSTETRICS.

A Practical Hand-book of Midwifery. By FRANCIS W. N. HAULTAIN, M.D. 18mo, profusely illustrated with original Cuts, Tables, etc.; about 260 pp., handsomely bound in leather, 6s.

A practical manual produced in a portable and convenient form for reference, and especially recommended for its compactness, conciseness and clearness.

"One of the best of its kind, and well fitted to perform the functions its author claims for it."—*Edinburgh Medical Journal.*

The Menopause and its Disorders. With Chapters on Menstruation. By A. D. LEITH NAPIER, M.D., M.R.C.P., late Editor of the *British Gynæcological Journal*, Author of *Notes on Puerperal Fever, The Thermometer in Obstetrics and Gynæcology*, etc., etc. Royal 8vo, cloth gilt, illustrated by a series of Original Photo-micrographs and many Illustrations in the Text, price 7s. 6d. net.

"The work is a monument of laborious research, and will always be referred to in the future as a standard authority on the subject."—*Practitioner.*

THE SCIENTIFIC PRESS, LTD., 28 AND 29 SOUTHAMPTON STREET, STRAND, W.C.

NURSING, TEXT-BOOKS ON.

Nursing: Its Theory and Practice. Being a complete Text-book of Medical, Surgical and Monthly Nursing. By PERCY G. LEWIS, M.D., M.R.C.S., L.S.A. New and Revised Edition (10th thousand), with entirely new chapters on Monthly Nursing and Confinements, profusely illustrated with over 100 new Cuts. Crown 8vo, cloth gilt, 3s. 6d.

To this edition have been added three chapters on Monthly Nursing, together with additions to the chapters on Blood Diseases, on Antiseptics, on the Nursing of Children, and on Private Nursing. Under these headings will be found a short account of the new Aseptic Treatment, the Antitoxin and Serum Treatments, and the Treatment by Thyroid Extracts.

These additions, combined with the revision throughout of the text, render this work the best and most complete Text-book on Nursing in the market, and strengthen its position as a Classic in the Training Schools of the World.

Elementary Anatomy and Surgery for Nurses. By WILLIAM MCADAM ECCLES, M.S., Lond., M.B., F.R.C.S., L.R.C.P., Surgeon, West London Hospital; late Senior Assistant Demonstrator of Anatomy, St. Bartholomew's Hospital, etc., etc. Crown 8vo, cloth, profusely illustrated, 2s. 6d.

"The book is well illustrated, and will be valuable to nurses who are beginning the study of anatomy. The instruction is given very clearly and concisely, and the reader is able to glean much information in a very condensed form."—*Nursing Notes.*

Ophthalmic Nursing. By SYDNEY STEPHENSON, M.B., F.R.C.S.E., Surgeon to the Ophthalmic School, Hanwell, W., etc., etc. A valuable contribution to Nursing literature on a subject hitherto never handled with special reference to Nursing. Crown 8vo, 200 pp., profusely illustrated with original drawings, list of Instruments required, and a Glossary of Terms, 3s. 6d.

"May very advantageously be studied by nurses and clinical clerks, or dressers who are about to enter the Ophthalmic service of a hospital."—*The Lancet.*

Elementary Physiology for Nurses. By C. F. MARSHALL, M.D., B.Sc., F.R.C.S., late Surgical Registrar and Anæsthetist to the Hospital for Sick Children, Great Ormond Street; formerly Platt Physiological Research Scholar in the Owen's College, Manchester. Crown 8vo, illustrated, cloth, 2s.

The Nurse's Dictionary of Medical Terms and Nursing Treatment. Compiled for the use of Nurses. By HONNOR MORTEN. Containing descriptions of the Principal Medical and Nursing Terms and Abbreviations, Instruments, Drugs, Diseases, Accidents, Treatments, Physiological Names, Operations, Foods, Appliances, etc., etc., encountered in the Ward or Sick-room. Third and Revised Edition (twenty-fifth thousand), demy 16mo (suitable for the apron pocket) handsomely bound in cloth boards, 160 pp., 2s., in handsome leather, gilt, 2s. 6d. net.

"A very useful little book for reference, and should be at the disposal of every nurse."—*Birmingham Medical Review.*

Mental Nursing. See p. 9.

THE SCIENTIFIC PRESS, LTD., 28 AND 29 SOUTHAMPTON STREET, STRAND, W.C

NURSING, TEXT-BOOKS ON—*(Continued)*.

Surgical Ward-work and Nursing. By ALEXANDER
MILES, M.D., Edin.; C.M., F.R.C.S.E. A practical Manual of
clinical instruction for Students in the Wards, and Nurses. Con-
cisely, simply, and comprehensively treated. Contents: Section I.—
Antiseptic Surgery, Section II.—The Use of Rest in Surgery.
Section III.—Bandaging. Section IV.—Surgical Instruments and
Appliances. Demy 8vo, cloth boards, copiously illustrated, 3s. 6d.

" An illustrated manual which will furnish much needed guidance to the
student and the nurse in the early stages of their apprenticeship."—*The Times.*

Nursing. By ISABEL ADAMS HAMPTON. Crown 8vo, 484 pp., 2nd
Edition, 7 Plates, 18 Illustrations, Charts, etc., 7s. 6d. net.
The Scientific Press have secured the English rights of this important work,
and a stock of copies is on sale at their premises. It is considered one of
the most important treatises on Nursing that has been published.

" It is not too much to say that in no single volume yet published has
the subject been so scientifically, carefully, and completely treated as it has
been by Miss Hampton. . . ."—*The Hospital.*

Practical Points in Nursing. For Nurses in Private Practice. With an Appendix containing Rules for Feeding the Sick ;
Recipes for Invalid Foods and Beverages; Weights and Measures ;
Dose List ; and a full Glossary of Medical Terms and Nursing Treatment. By EMILY A. M. STONEY. 12mo, illustrated with 73 En.
gravings in the Text, and 9 coloured and half-tone Plates. Price
7s. 6d. NET.

" The hints given are excellent, and we hope they may be widely read."
British Medical Journal.

Art of Massage. By A. CREIGHTON HALE. Demy 8vo, 200 pp.,
cloth gilt, profusely illustrated with over 70 special Cuts, 6s.

" Mrs. Hale seems to have had considerable experience not only as a
practitioner, but as a teacher of the art, and explains in a series of clearly
illustrated articles the many complaints for which massage has been found
beneficial, and the special manipulations required in each case."—*Morning
Post.*

" A volume on the ' Art of Massage,' from the pen of one of its foremost
advocates in this county. . . . Admirably illustrated."—*Westminster Review.*

" The latest and most complete book we know on Massage."—*Queen.*

Pamphlets on Nursing.

No 1. The Advantages and Privileges of a Trained District Nurse, and
the Duties of the District towards Her. By HENRY C. BURDETT.
1½d. per copy, or 1s. per dozen.

" A pamphlet advocating the employment of none but thoroughly trained
persons as district nurses."—*Essex Herald.*

No. 2. A Great Movement—The Nurses' Co-operation. By Mrs.
FURLEY SMITH. Price as No. 1.

" Those who are nurses or about to become nurses will find something
worth considering in this little pamphlet. A sketch of the work done is
given, with rules for any similar association that might be started."—*Kendal
Mercury and Times.*

THE SCIENTIFIC PRESS, LTD., 28 AND 29 SOUTHAMPTON STREET, STRAND, W.C

NURSING, TEXT-BOOKS ON—(*Continued*).

How to Become a Nurse: and How to Succeed. A complete guide to the Nursing Profession for those who wish to become Nurses, and a useful book of Reference for Nurses who have completed their training and seek employment. Compiled by HONNOR MORTEN. Author of "The Nurse's Dictionary," etc. Third and Revised Edition, Demy 8vo, 200 pp., profusely illustrated with Copyright Portraits and Drawings, 2s. 6d.

Contains Alphabetical List of the Chief Institutions in the United Kingdom and Abroad where nurses are trained and employed; Fees; Qualifications and all particulars; information as to Application, Probation, Certification, Midwifery, etc.; Private, Infirmary, District, Asylum, Army and Navy, and Nursing Abroad; Male Nursing; Massage; Case-taking; Uniforms (Illustrated); Matrons; Nursing Societies and Guilds; Equipment; Examination Questions, etc., etc.; and LIVES OF SOME EMINENT NURSES (with Copyright Portrait of each).

"To those who are frequently appealed to by girls in their teens, or by young women of mature years as to the steps they should take to become nurses, **this book of Miss Morten's must prove a perfect godsend.** It deals with the nursing arrangements of most, if not all the training schools in London and throughout the country, and gives the novice a thorough insight into the duties expected from her. To those entering, or desirous of entering, any branch of the profession, there is no other guide we know of which contains in so small a compass more useful or practical information."— *British Medical Journal.*

Hospital Sisters and their Duties. By EVA C. E. LÜCKES, Matron to the London Hospital. Third Edition, enlarged and partly re-written, crown 8vo, cloth gilt, about 200 pp., 2s. 6d.

The experience and position of its author entitle it to authority, and every Matron, Sister or Nurse who aspires to become a Sister should read the advice and information contained in its pages.

"An admirable guide to the beginner anxious to take up a Sister's work. Its pages bring help and knowledge imparted in a very sympathetic, straightforward manner by one who has had practical experience in the lesson she teaches."—*Gentlewoman.*

Ministering Women: The Story of the Royal National Pension Fund for Nurses. By GEORGE W. POTTER, M.D. Demy 8vo, cloth, 150 pp., profusely illustrated, 2s. 6d.

The book not only deals with the Royal National Pension Fund and the history of its foundation and development, but it also deals with a number of subjects of special interest to nurses.

"Should be widely circulated among trained nurses and their friends."— *Record.*

"The Hospital" Nurse's Case Book. For use in Hospital as well as Private Nursing. Demy 16mo (for the apron pocket), 72 pp., strong boards, Second Edition, 6d., or 4s. 6d. per dozen (post free).

A Book of Tables specially prepared for use by Nurses in the Ward and in the Sick-Room. Containing Twenty-Six complete Tables, each being ruled to last one week, for keeping an exact record of a Patient's condition, including notes on Temperature, Pulse, Respiration, Motions, Age, Sex, Name, Address, Occupation, Treatment, History, Name of Medical Attendant, Number of Case, and Date of Admission.

THE SCIENTIFIC PRESS, LTD., 28 AND 29 SOUTHAMPTON STREET, STRAND, W.C

16

PERIODICALS.

Science Progress: a Quarterly Review of Current Scientific Investigation. Conducted by Sir HENRY C. BURDETT, K.C.B., and Edited by J. BRETLAND FARMER, M.A., with the Co-operation of a powerful Editorial Committee. Price 3s., or 3s. 3d. post free. Subscription Price, 10s. 6d. per annum, post free.

" Has now entered well into its stride as the representative of advanced science."—*Pall Mall Gazette.*

" Each of the papers in *Science Progress* is an important addition to scientific literature. The contributors to this monthly review of current investigations are always men in thorough touch with their subjects, and the result is that they summarise all that is worth knowing on the subjects treated by them."—*Nature.*

" Such an influential and eminent editorial authority command support and inspire confidence ; the support has been readily given, and the confidence fully justified. To be fully in touch with current scientific investigation we must read *Science Progress.*"—*Science and Art of Mining.*

(*A Special American Edition is published by Messrs. D. C. Heath &*
Co., of Boston, Chicago and New York.)

The Hospital; A Journal of the Medical Sciences and Hospital Administration, with which is issued Weekly a SPECIAL SECTION FOR NURSES.

The Hospital is read everywhere, is quoted everywhere, and is accepted everywhere as the medical journal which is the most truly representative of modern medical science.

TERMS OF SUBSCRIPTION:
PRICE TWOPENCE.

Published every Saturday.		*Eleventh Year of Publication.*
For the United Kingdom,	.. ..	10s. 6d. per annum, post free.
For the Colonies and Abroad, ..	..	15s. 2d. ,, ,,
Or may be had in Monthly Parts,	..	8s. 6d. ,, ,,

Subscriptions may commence at any time.

Burdett's Hospitals and Charities. Being the Year Book of Philanthropy. Containing a Review of the Position and Requirements, and Chapters on the Management, Revenue and Cost of the Charities. An exhaustive Record of Hospital Work for the Year. It will also be found to be the most Useful and Reliable Guide to British, American, and Colonial Hospitals and Asylums, Medical Schools and Colleges, Religious and Benevolent Institutions, Dispensaries, Nursing and Convalescent Institutions. Edited by Sir HENRY C. BURDETT, K.C.B. Published annually, 1000 pp., crown 8vo, cloth gilt, 5s.

" This invaluable manual contains a quite exhaustive statement of its subject. What hospitals do, for whom they do it, by whose help and at whose cost, are every one of them questions of the profoundest interest, and they find their answer here."—*Spectator.*

" Still remains the standard work of reference upon all points which relate to charities, British, American, and Colonial."—*Daily Telegraph.*

" This book tells us more about hospital work, medical colleges, etc., than any book we have ever seen in print. It contains more matter, more figures, more correct data that people know nothing of, than any book ever written."—*New York Medical Journal.*

Burdett's Official Nursing Directory. Compiled and Edited, with the assistance of a Committee of Medical Men and Matrons, by Sir HENRY C. BURDETT, K.C.B. Will contain the fullest and most complete list of Nurses and Nursing Institutions ever issued. Published annually, price 5s. [*Shortly.*

THE SCIENTIFIC PRESS, LTD., 28 AND 29 SOUTHAMPTON STREET, STRAND, W.C.